Health Research Design and Methodology

Editor
Eucharia Nnadi Okolo, Ph.D.
Professor
College of Pharmacy and Pharmaceutical Sciences
Florida A&M University
Tallahassee, Florida

CRC Press
Boca Raton Ann Arbor Boston

Library of Congress Cataloging-in-Publication Data

Health research design and methodology/editor, Eucharia Nnadi Okolo.
p. cm.
Includes bibliographical references.
Includes index.
ISBN 0-8493-4627-4
1. Medical sciences—Research. 2. Research management.
3. Medical writing. 4. Medical statistics. I. Okolo, Eucharia
Nnadi, 1954-
[DNLM: 1. Data Colection—methods. 2. Data Interpretation,
Statistical. 3. Research Design. 4. Writing. W 20.5 H4336]
R850.H38 1990
610.72—dc20
DNLM/DLC
for Library of Congress 90-1953
CIP

Direct all inquiries to CRC Press, Inc., 2000 Corporate Blvd., N.W., Boca Raton, Florida 33431.

International Standard Book Number 0-8493-4627-4

Library of Congress Card Number 90-1953
Printed in the United States

DEDICATION

This book is dedicated to all those with inspiring and inquiring minds, thus leading to improvements and new developments. To all my teachers and inspirators who taught me that limits are what I make them. To my mother, Lucy James Oguanobi, who has always encouraged me to reach out and make each new day an improvement on the day before. To all those who have helped in the process of making this book a reality.

Eucharia Nnadi Okolo

PREFACE

''Health Research Design and Methodology'' covers various research design methodologies and analyses in the health sciences. Emphasis will be placed on appropriate application of the mentioned topic areas rather than on theory. The book contains examples of research design, methodology, and analysis in the health sciences. The book, in addition, focuses on (1) proposal writing — ie., the key elements of a research proposal; (2) interpretation of research data based on analysis, and (3) reporting research findings.

The book is designed especially for:

1. Students in the area of health sciences who are involved in writing research proposals and conducting research
2. Health science research instructors
3. New researchers in the various areas of health sciences

Research is a vital part of health sciences, especially in the areas of education and industry. Various advanced health sciences programs do require successful completion of research work as a prerequisite to obtaining a degree. The success of any health science research depends on appropriate design, methodology, analysis, and interpretation. ''Health Research Design and Methodology'' will therefore play a vital role in aiding reseachers — especially student researchers, instructors, and new researchers — in their various areas of the health sciences.

Eucharia Nnadi Okolo
Tallahassee, Florida
July 1990

THE EDITOR

Eucharia Nnadi Okolo, Ph.D., is a Professor of Pharmacy Administration in the College of Pharmacy and Pharmaceutical Sciences at Florida A&M University, Tallahassee, Florida. She received her B.S. in Pharmacy from Creighton University, Omaha, Nebraska in 1977, graduating with honors, and less than 1 year later completed her Master's degree in Hospital Pharmacy at the University of Minnesota. In 1982 she received her Ph.D. degree from the University of Minnesota, Minneapolis.

Since 1981 she has held successive appointments as Assistant Professor, Associate Professor, and currently Professor in the College of Pharmacy and Pharmaceutical Sciences at Florida A&M University, Tallahassee, Florida.

Dr. Nnadi Okolo has taught several courses including Biostatistics and Research Methodologies to both graduate and undergraduate students. She is active in various research studies and grants, including international research activities. Dr. Nnadi Okolo is the author of several publications and has refereed professional papers; she is an editorial board member of *Clinical Research Practices and Drug Regulatory Affairs Journal*. She is very active in several College and University committees, and she has served as a member of the University Senate. Dr. Eucharia Nnadi Okolo is a member of Rho Chi (Mu Chapter), the Professional Pharmaceutical Honor Society, the American Association of the Colleges of Pharmacy, and the American Public Health Association.

CONTRIBUTORS

Collins O. Airhihenbuwa, Ph.D.
Assistant Professor
Pennsylvania State University
University Park, Pennsylvania

Johnny L. Early, Ph.D.
Professor and Dean
College of Pharmacy and Pharmaceutical Sciences
Florida A&M University
Tallahassee, Florida

Jack E. Fincham, Ph.D.
Associate Professor
School of Pharmacy
Samford University
Birmingham, Alabama

Abraham G. Hartzema, Ph.D.
Associate Professor
School of Pharmacy
University of North Carolina
Chapel Hill, North Carolina

Peter D. Hurd, Ph.D.
Associate Professor
St. Louis College of Pharmacy
St. Louis, Missouri

Arcelia Johnson-Fannin, Pharm.D.
Associate Professor
College of Pharmacy and Pharmaceutical Sciences
Florida A&M University
Tallahassee, Florida

Hugh F. Kabat, Ph.D.
Professor
College of Pharmacy
University of New Mexico
Albuquerque, New Mexico

Bryan A. Lewis, Ph.D.
Associate Professor
College of Pharmacy
Florida A&M University
Tallahassee, Florida

Michael Montagne, Ph.D.
Associate Professor
Department of Pharmacy Practice
Northeastern University
Boston, Massachusetts

Eucharia Nnadi Okolo, Ph.D.
Professor
College of Pharmacy and Pharmaceutical Sciences
Florida A&M University
Tallahassee, Florida

Knife Redda, Ph.D.
Professor
College of Pharmacy and Pharmaceutical Sciences
Florida A&M University
Tallahassee, Florida

Charles A. Walker, Ph.D.
Chancellor
University of Arkansas at Pine Bluff
Pine Bluff, Arkansas

MinQi Wang, Ph.D.
Assistant Professor
Pennsylvania State University
University Park, Pennsylvania

TABLE OF CONTENTS

Chapter 1

INTRODUCTION TO WRITING A RESEARCH PROPOSAL

E. Nnadi Okolo, H. F. Kabat, J. L. Early, K. Redda, A. Johnson-Fannin, C. A. Walker, and B. A. Lewis

TABLE OF CONTENTS

* E. Nnadi Okolo.

** A. Johnson-Fannin.

***H. F. Kabat.

† K. Redda.

I. INTRODUCTION

One of the major problems facing several students and faculty members in various health science areas involves writing research proposals, getting them approved and/or funded, and actually carrying out the research. For the student, writing a research proposal as a part of fulfilling a degree requirement can be a source of anxiety and uncertainty. Most schools do not require that a student research project be funded externally before approving the project for the student. However, it is expected that faculty receive external research funding. External sources include any source of funding outside of the institution, such as government agencies, various foundations and corporations, voluntary agencies, and organizations. Some of these external sources have specific areas of interest and will therefore only fund research projects addressing their interests and needs. In such instances, the researcher may not face the task of deciding on a research topic. Where the funding source has no specific area of interest, the researcher must decide on a specific topic based on his/her interest and expertise. This chapter will therefore guide an individual through the process of deciding on a research topic, identifying the proposal format, funding the research proposal, the process of proposal

* J. L. Early and C. A. Walker. Sponsored by NIH/DRR/RCMI (RRO 3020) and NIH/DRR/MBRS (RR) 8111.

review, and learning how to increase the probability of funding a research proposal; examples of funded and unfunded research projects are included.

II. DECIDING ON A RESEARCH TOPIC

A. INTRODUCTION

If you know what subject you want to research, consider yourself lucky. You may be able to proceed directly to the workbench. If you have not selected a research topic, you might find that getting started is more difficult than imagined. The information contained in this section should provide the push to get you off square one and into an investigative effort.

B. DISCOVERING YOUR INTERESTS

Discovering your interest involves several steps, such as self-examination, examination of prior projects, delineation of future goals, and generation of an interest list.

1. Self-Examination

When one embarks on a research track, it is likely that this path will produce a long association. New projects will probably be tangential or cross-related to the original idea. Therefore, the initial decision of what to study should include an examination of feelings about research. If, for example, the researcher cannot tolerate white mice, faints when blood is drawn, or has deep-seated beliefs about using dogs in experiments, then projects which include or cause such should be avoided. If the application of highly sophisticated chemical assay techniques to serum provokes a high degree of satisfaction, this might be an appropriate research arena.

The potential investigator should take time to examine physical and mental limitations. If it is apparent that a research effort will require long walking forages into the woods to gather samples, the physically impaired person or asthmatic, for example, might want to avoid such research arenas.

2. Examine Prior Projects

Most college-level students have done some sort of investigation. Recall that paper written for advanced biology, the zoology group project, and the scientific work done for an R0 1 grant. List any projects, papers, or activities which can be called "investigative". Expand the list by recalling what was good about each, what could be changed, and if it is worthy of further pursuit.

3. Delineate Future Goals

Discovering a method for separating and identifying the numerous metabolites of an antipsychotic will require years in a laboratory setting. Discovering a correlation between serum level of that same antipsychotic and patient response will necessitate involvement in clinical situations. The decision to pursue one or the other determines whether the researcher becomes primarily a scientist or a clinician. It is important to define goals for the long term as well as the short term.

4. Generate an "Interest List"

With the long- and short-range goals clearly visible and the prior works list at hand, make a list of anything which is of interest and appears to fit. Do not pay attention at this point to the scope or syntax. Write what comes to mind. Eliminate anything which crept in that is not in keeping with physical and emotional limitations. At this point, do not do any other editing. This is a dynamic list of topics which stimulate the intellect and appeal to curiosity. It will change in character and eventually become a viable pot from which to draw future research.

C. SEARCHING THE LITERATURE

1. Necessity for a Search

If unlimited time and resources were available, perhaps all topics on the "interest list" could be researched. However, since this is rarely the case, choices must be made. A mandatory step toward making the choice is to examine the literature for information relative to the potential topic. Science grows by adding new information on top of the foundations already formed. A worthwhile new contribution cannot be made if one is ignorant of what was known before.

Therefore, the two goals of a literature search are (1) find out if the objective of the proposed project has already been adequately explained by prior works and (2) acquire a broad general background of the subject matter. Both goals usually can be accomplished by reading a small but carefully selected segment of available information.

2. How to Search

Access to a good library is essential. The library would be expected to have available the standard references, textbooks, periodicals, and abstracting services. However, to keep abreast with today's information fast track a search through the periodical literature is necessary. It is not prudent, however, to plunge into the latest volume of *International Pharmaceutical Abstracts* (IPA) with the entire "interest list". Some preliminary editing must be done. Group together any subject matter which appear to be related. Delete any headings which are no longer appealing. Choose the four to six which arouse the most curiosity. Now set some limits for the search — what period it will cover, what sources (see Section II.C.2.b below) will be searched, and any other desired limitations.

a. Structure of the Periodical Literature

The periodical literature can be searched manually from print sources or by computer (automated) *print*. The indexing services have made it easy to review what has been printed in thousands of journals and other publications.

The major indexing and abstracting services for scientific literature include the following:

- Index Medicus — cites articles drawn from over 2500 medical and paramedical journals worldwide.
- Excerpta Medica — an abstracting journal arranged into sections scanning some 3,500 journals.
- International Pharmaceutical Abstracts (IPA) — a minor (by comparison) but important service for pharmacy-oriented information. Abstracts culled from hundreds of pharmacy and related titles are presented.
- Biological Abstracts — a vast abstracting service which comprehensively covers the bioscientific literature.
- Chemical Abstracts — an abstracting service of medically related secondary source literature directed toward chemistry and chemistry technology.
- Science Citation Index — a unique interdisciplinary system which provides notation of authors who have cited earlier works. There are several interrelated indexes in the system which can be used together to create a bibliography and expand the scope of materials.

b. Search Procedures and Sources

A literature search can be done manually or through automated means.

TABLE I
MEDLARS And DIALOG Files

MEDLARS	DIALOG
	Medicine and Bioscience
Avline	Biosis Previews (1969—present)
Bioethics	CA Search (1967—present)
Cancerlit	Chemname
CancerProj	Chemsearch
Catline	Chemsis (1967—present)
Clinprot	Chemzero
Chemline	Embase (Excerta Medica)
Health	Health Planning
Histline	IPA
Medline (Index Medicus)	Life Sciences Collection
MeSH (vocubulary)	Medline (1966—present)
Popline	Medical Health Abstracts
RTECS	Pharmaceutical News Index
Serline	SCISearch (1974—present)
Toxline	Telegen

TABLE II
Selected ISI/BIOMED Headings

Allergy	Medicine
Anatomy and morphology	Microbiology
Anesthesiology	Molecular biology
Biochemistry	Neurosciences
Biology	Nutrition
Biometrics	Obstetrics and gynecology
Cancer	Orthopedics
Cardiovascular system	Pediatrics
Dermatology	Pharmacology and pharmacy
Drugs and addiction	Psychiatry
Endocrinology	Rheumatology
Gastroenterology	Surgery
Genetics	Toxicology
Geriatrics	Urology
Immunology	Virology

Manual Searching — While a degree of satisfaction may be obtained from manual searching, it is a time-consuming process. There is also the possibility of overlooking important information. The automated or on-line resources can help circumvent these problems. There is a fee attached to on-line searching.

Automated (On-Line) Sources — Innumerable data bases are available for automated search. The creators of data bases lease their data bases to search services (hosts). Two of the better known and more widely used hosts are MEDLARS and DIALOG. MEDLARS hosts a variety of files (see Table I) primarily of interest to the medical and scientific community. Index Medicus as MEDLINE is a primary data base. DIALOG hosts more than 190 data bases covering a wide range of subjects (e.g., Index Medicus and Excerpta Medica).

Another on-line system supplier, the Institute for Scientific Information (ISI), hosts its own files — the Science Citation Index. The data base, ISI/Biomed, is of practical use to medically related researchers. (See Table II.)

c. *Heading for Search*

With the help of a good librarian, a manual or on-line search can be accomplished

TABLE III
Sample MeSH Heading

PSYCHOTHERAPY
F4.754+
XU GESTALT THERAPY
XU PSYCHOTHERAPY, RATIONAL-EMOTIVE

PSYCHOTHERAPY, BRIEF
F4.754.738
73
X SHORT TERM PSYCHOTHERAPY

PSYCHOTHERAPY, GROUP
F4.754.864.581+
X GROUP THERAPY

PSYCHOTHERAPY, MULTIPLE
F4.754.766
73

PSYCHOTHERAPY, RATIONAL see PSYCHOTHERAPY, RATIONAL-EMOTIVE

PSYCHOTHERAPY, RATIONAL-EMOTIVE see under PSYCHOTHERAPY
X PSYCHOTHERAPY, RATIONAL
X RATIONAL-EMOTIVE PSYCHOTHERAPY

PSYCHOTIC DISORDERS
F3.709.680+
81; was PSYCHOSES 1963-80
X PSYCHOSES
X PSYCHOSIS, BRIEF REACTIVE
X SCHIZOAFFECTIVE DISORDER
X SCHIZOPHRENIFORM DISORDERS
XU CAPGRAS SYNDROME

PSYCHOTOMIMETIC AGENTS see HALLUCINOGENS

PSYCHOTROPIC DRUGS
D15.236+
75; was see under ANTIDEPRESSIVE AGENTS & TRANQUILIZING AGENTS 1974
see related
PSYCHOPHARMACOLOGY
X PSYCHOACTIVE AGENTS
X PSYCHOPHARMACEUTICALS
XR PSYCHOPHARMACOLOGY

Note: Citation will be found in Index Medicus only for "bold type" headings. On-line data bases may be searched using all "see under" cross-references. X = "see", XU = "see under", XR = "see related".

From *Cumulated Index Medicus, Vol. 29, 1988.*

easily. Before approaching the information files, however, appropriate headings must be determined. A good reference for headings is the Medical Subjects Headings (MeSH) of Index Medicus. Take the topics selected from the "interest list" to the MeSH compilation. This is a good first step even if an on-line search is planned. Generate a list of search headings for each topic by identifying the MeSH heading, cross-references, and other closely related terms. (See Table III.) The librarian can assist with defining the specifics of on-line searching.

3. Organization of Information

Although this search is undertaken to gather preliminary information, remember to keep good records. Make complete bibliography notations. Make notes of abstracts and articles reviewed. Add to and subtract from the "interest list". Keep materials for each subject together.

D. CULTIVATING THE TOPIC

The "interest list" has changed. Two topics have been eliminated, others added, and some refined. The preliminary forage into the literature may have revealed that a significant amount of work has been done to develop a chemical assay for the amount of an antipsychotic drug in the serum. All efforts were less than satisfactory for the same reason. If that limiting factor is not yet surmountable, discovering this chemical assay might be insurmountable. This particular topic would then have to be eliminated or restated to reflect the information provided by the literature. This exercise should be done with all remaining topics.

The literature should have provided information to help make a well-informed decision about the requirements to conduct a particular investigation. For each topic, answer the questions:

- How much time will it require?
- Can that much time be devoted to this?
- What space will be needed?
- Will adequate funds be available, if needed?
- Will trained auxiliary personnel be needed?
- What specialized equipment is required?

A critical application of these points to the ''interest list'' will define it further. Now that all of the logical steps have been applied, rank the topics based solely on feelings. Of the remaining topics, which one has the most appeal? At this point, a consultation with established researchers should help bring out any aspects of the proposed investigations which need to be considered.

E. CHOOSE THE FINAL TOPIC

Now it is time to state the problem correctly. Sometimes the problem which is enumerated is not what the researcher actually intends to do. Grammar is important. Syntax is important. If the intent is to look for metabolites of that antipsychotic drug in pooled urine, then do not say ''pooled urine metabolites''; say ''metabolites in pooled urine''. Take the time to express the direction of the research in a concise descriptive statement. It has been said that many scientists owe their greatness not to their skill in solving problems, but to their wisdom in choosing them.

REFERENCES — SECTION II

1. **Day, R.,** *How to Write and Publish a Scientific Paper,* 2nd ed., isi Press, Philadelphia, 1983.
2. **Wood, S. M., Horak, E. B., and Snow, B.,** *End User Searching in the Health Sciences,* Haworth Press, New York, 1986.
3. **Harter, S.,** *Online Information Retrieval: Concepts, Principles, and Techniques,* Academic Press, Orlando, FL, 1986.
4. **Welch, J. and King, T.,** *Searching the Medical Literature: A Guide to Printed and Online Sources,* Chapman and Hall, New York, 1986.
5. **Huth, E.,** *How to Write and Publish Papers in the Medical Sciences,* isi Press, Philadelphia, 1983.
6. **Wilson, E., Jr.,** *An Introduction to Scientific Research,* McGraw-Hill, New York, 1952.
7. **Noland, R.,** *Research and Report Writing in the Behavioral Sciences,* Charles C Thomas, Springfield, IL, 1970.

III. KEY ELEMENTS IN WRITING A RESEARCH PROPOSAL

While it is possible to conduct health research in which funding is not a primary factor, most health research is dependent upon funding from sources where there is competition for scarce funds. The most competitive proposals are those which address the following key elements prior to approaching the potential funding sources:

1. *Conduct a literature search.* An exhaustive bibliographic search must be conducted at an early stage of proposal development in order to avoid duplication of existing studies and to identify studies which may support the work through citations in the proposal presentation. This search can also include contacts with investigators engaged in similar research.
2. *Consultation with colleagues.* It is frequently productive to consult with colleagues about a proposal throughout the development stage. They may be able to suggest approaches to use, ways to strengthen the proposal, and possible funding sources.
3. *Assess your personal qualifications and the proposal's requirements.* It is essential to assess one's own ability to execute the proposal. This assessment must include a determination of the capabilities and expertise required by the proposal, but should also consider essential questions of availability of staff, equipment, time, and other institutional support including financial commitments. The particular talents and other support needed to accomplish the proposal must be built into the proposal from the beginning.
4. *Assess the financial scope of the proposal.* The principal investigator should determine the financial scope of the proposal at an early stage to determine whether the proposal's support needs can be met by the funding source.
5. *Identify the needs and priority interests of potential funding sources.* Initially, unless you are responding to a governmental notice of program intent (NPI), a request for proposal (RFP), or a request for quotation (REQ), it is advisible to research the funding source to determine its general purpose and activities, areas of interest, and previous grant disbursements. You may also contact the funding source's staff members to determine their interest in the proposal. This can be accomplished by a telephone conversation, office visit, submission of a letter of intent, or any combination of these. A letter of intent is an introductory proposal for the potential funding source and serves as the basis for a subsequent preliminary or full proposal. The letter of intent should be limited to approximately one or two pages in which the major purpose and need for the proposal, objectives, and methodology are outlined. This letter should be succinct and include only necessary information (no supportive documentation). The letter of intent enables the funding sources to screen unsolicited proposals which may be inappropriate for their current areas of interest. The response from the funding source may well provide the investigator with valuable information to further develop or redirect the proposal.

A. PROPOSAL FORMAT

In order for good ideas to be competitive in the funding process, it is essential that a well-developed and clearly defined proposal be submitted to the funding source for review and consideration. The proposal format should always adhere to the guidelines of the funding source. Most governmental agencies and private/corporate foundations are receptive to proposals that follow a "standard" format with which their proposal reviewers are relatively familiar. Whether or not there are specific guidelines and application forms, there are certain components required in all proposals. These components are as follows:

- *Cover sheet* — Always use the form suggested by the funding source. If one is not specified, then a cover sheet should include the name and complete address of the prospective funding source; the name and complete address of your institution; the title of the proposal, period of performance, and amount requested; the investigator's name, signature, title, social security number, and telephone number; and the name, signature, title, telephone number, and date of each person in your organization who must sign proposals.

- *Abstract* — A one-page condensation or overview of the proposal should immediately follow the cover sheet. The abstract should cover the essential elements of the proposal. The abstract is used by the reviewers as an overview of the proposal before they examine its individual components and is sometimes used by the funding source to disseminate the proposal for comments from individuals outside the immediate primary review group.
- *Table of contents* — A table of contents is generally included in all proposals more than five pages in length. This is a listing of page numbers for all chapters, sections, subsections, or other parts of the proposal in the order in which they appear.
- *Introduction* — The introduction includes (1) enough introductory material to make possible an understanding of the problem, (2) an explanation of the assumptions and definitions of terms, and (3) a one-sentence statement of the problem.
- *Narrative* — The narrative is a description of the proposal which details precisely what will be done.
- *Background* — This section details what has been accomplished to date, what requires attention, and what requires additional attention or refocusing.
- *Statement of need* — This is a clear and concise statement of what is to be done and why.
- *Proposal goals* — These are generalized statements which detail proposed accomplishments.
- *Proposal objectives* — These are statements of precise and measurable outcomes.
- *Proposal implications* — The tentative results and impact of the proposal addressing the questions of what and why are detailed.
- *Literature review* — Your familiarity with similar studies, either in progress or completed, as well as complementary studies which may impact on the proposal is cited. You should indicate whether or not you have been involved in similar work.
- *Methodology* — This is an explanation of how the problem will be approached and what methods will be used to accomplish the stated objectives. A solution methodology is provided for each stated objective. Specific descriptions of actual program activity, personnel, and data compilation are included. If statistical techniques are to be employed, they should be detailed. If a sample is to be drawn, methods of selection and basic information concerning the sample are needed.
- *Time frame* — A projected time frame with reasonable starting and ending dates for the proposal must be included. Provide a proposal flow chart indicating anticipated accomplishments at specific times.
- *Dissemination (utilization)* — Detail how findings will be disseminated or utilized. Provide a plan for utilization or transferability of the findings.
- *Facilities* —This includes a statement of the physical facilities, library, data processing, and space available and necessary to accomplish the proposal.
- *Personnel* — Describe the relationship of the professional personnel to the proposed study. Attach vitae of the principal proposal personnel in the appendices.
- *Budget* — Develop a line-item budget to support the proposal. Provide a separate budget explanation or justification to assist the funding source in understanding the budget.
- *Appendices* — This section includes (1) an orderly list of bibliographical citations for all literature cited; and (2) vitae for the senior proposal personnel, which include a list of publications and previously funded projects.

IV. FUNDING OF A RESEARCH PROPOSAL

In the U.S., several research grant-awarding institutions in the scientific field can be identified. Among the most prominent agencies are the National Institutes of Health (NIH), the National Science Foundation (NSF), the National Aeronautics and Space Administration (NASA), the Department of Defense (DOD), the Department of Agriculture (DOA), the Department of Energy (DOE), numerous foundations, several state governments, and a few industrial and pharmaceutical firms. The overwhelming majority of people who benefit from these grant awards serve the accredited U.S. colleges or universities and nonprofit research institutes.

Most research grant awarding agencies periodically make specific program announcements and request the submission of grant applications within a specified period. The Federal Register, for example, compiles announcements concerning applications for grants sponsored by federal agencies. Moreover, each organization prepares grant application forms with detailed instructions, and the immediate task of writing a research proposal is familiarizing oneself with these forms and the set of isntructions. NIH, within the Public Health Service (PHS) of the U.S. Department of Health and Human Services, uses the grant application form and instruction sheets of PHS 398 as its official document.[1,2]

Interested individuals or grant-receiving institutions can request the shared instrument grant, individual or institutional research (and research training) grant, animal resources grant, resources development (resource-related project) grant, library resource improvement grant, newcomer initiation (First Independent Research Support and Transition) grant, faculty development grant, program project and center grant, publication grant, and pre- and post-doctoral fellowship grant.[1,3]

For beginning researchers, the most difficult part of doing research is finding a funding source. The Foundation Directory, published biennially, contains the names and purposes of large private foundations. The Foundation Grants Index, published annually in a book form and bimonthly otherwise, gives an indication of donor, recipient, and subject matter funded. Its subject index helps identify the right foundation that potentially might be interested in the project. The Directory of Research Grants, published annually, is another fine source, since it is indexed by subject matter as well as funding source. A major potential source to keep track of is the Catalog of Federal Domestic Assistance. From it one can identify programs, addresses, and names of contact persons for such federal agencies as the NIH and the NSF.[2]

Funding of a research proposal strictly depends on the results of a review by a panel of experts selected to examine the project — for example, the submitted proposal is reviewed based on the merit of its specific aim, its background, methodology (hypothesis and study design), analysis of data and conclusions, significance of study, expertise of the principal investigator and the research team, budget, and resources and environment.[4-7]

A well-written proposal is normally the first step toward funding a project or study. However, in their eagerness to transform their unique idea into a proposal, inexperienced researchers can easily overlook many routine details. Their time calculations may become faulty, and the deadline for submission approaches before they can finish. Their proposals may remain ideas whose time has come and gone. It becomes essential to establish a computer file, sometimes referred to as the ''swipe file''. This is a file containing all the routine essentials for successful proposal writing except the idea. These essential components include updated biographical sketchs of the research team, formulas or formats for budget outlines (e.g., percent of fringe benefits and indirect costs), organizational structure, available environment, and other research support.[2]

V. RESEARCH PROPOSAL REVIEW[8]

Every grant-funding agency has in place an Initial Review Group (IRG), a committee of experts that reviews submitted research proposals. The committee conducts a thorough analysis of the strengths and weaknesses of the proposals, assigns priority scores, and makes appropriate recommendations based on the scientific and technical merit of the proposals. Often, primary, secondary, and tertiary reviewers are selected, assigned to make an in-depth study of a given proposal, and asked to submit a written critique to the review committee to generate a thorough discussion on the proposal's scientific and technical merit.

In most cases, the following review criteria are considered in the evaluation of a research proposal.

A. IDENTIFICATION

Every proposal must contain an appropriate title, usually not more than 60 characters (spaces). It should clearly indicate the full names, addresses, and social security numbers of the principal investigator(s)/program director as well as the names and titles of the research team. Usually, the funding agency will assign an application number to each proposal for further identification.

B. GRANT APPLICATION HISTORY

The review committee prepares a brief description of the administrative history of the proposal, i.e., the renewal, supplemental, amended, resubmitted, or deferred application. If a prior review has been conducted, the review date and the contents of the committee's recommendations will have to be specified. If site visits have been made in the past, appropriate reports will be indicated. If there is no prior history, the ''New Proposal'' will be simply stated.

C. DESCRIPTION

This section concisely summarizes the body of the proposal. The review committee expects answers for the following questions: What does the research intend to do? What is the significance of the investigation? What has been known in the past about the particular study (literature survey)? Have there been any pilot studies by the principal investigator? How will the proposal work be accomplished? How appropriate and scientifically sound are the experimental protocols? What kind of data will be collected, and what conclusion can be drawn from the investigation? The description should be more that a restatement of the investigator's abstract and is prepared by the primary reviewers.

The issues that must be addressed are described in the following sections.

1. Specific Aims

The review committee briefly describes what the research is intended to accomplish and/or what hypotheses are to be tested. In situations where the hypotheses may be structured less formally or are absent (e.g., as in a descriptive or case study) the research questions to be addressed are stated.

2. Background

The primary reviewers highlight the background of the project based on literature cited and the previous experiences or preliminary studies of the investigator. If it is a competing renewal or supplemental application, the work accomplished in the previous grant period is summarized. The committee indentifies gaps which the project is intended to fill and states the importance of the proposed research by telling how the specific aims relate to the long-term objectives.

3. Methodology

The applicant's experimental design and procedures for accomplishing the specific aims of the project are described concisely. Included in the description are the number and type of subjects to be involved and how the sample will be chosen; a description of the control/comparison groups; the independent, dependent, and confounding variables; protocols or instruments to be used; plans for data collection and analysis; and any new methodology and its stated advantage over existing methodologies.

4. Collaborative Arrangements/Contracts

Any collaborative arrangements, cooperative agreements, or contracts with other parties, agencies, institutions, or firms which the applicant proposes to enter into in order to accomplish the project objectives are described briefly by the IRG.

D. HUMAN/ANIMAL SUBJECTS

1. Animal Subjects

Whether the research makes use of warm-blooded animals is noted. If so, whether the experimental procedures are acceptable and the animal care adequate is stated. The species of all animals used in the study are identified.

2. Human Subjects

When an exemption to human subjects regulations is claimed by an applicant, the committee must determine whether the claim is valid. If such an exemption is not applicable, then the subject population to be studied and its characteristics (e.g., anticipated number, source, age, sex, ethnicity, and state of health) will be described. The research project should specify the criteria for inclusion or exclusion and the rationale for the use of special classes of subjects, i.e., fetuses, children, institutionalized, etc. The review committee examines how subjects are to be recruited. Next, the committee gives its assessment of the necessity for using the subject population to be studied and the likelihood and nature of any potential risks — physical, psychological, social, legal, or other — to which the subjects might be exposed (e.g., invasion of privacy, legal consequences, coercion to participate, physical stress or injury). The adequacy of procedures to protect the rights and welfare of subjects, including an assessment of consent procedures (completeness of information given to subjects, how and from whom consent is obtained, the necessity of foregoing formal/informal consent, and a mechanism to replace it), confidentiality of data (how will records be safeguarded?), and other procedures such as the screening and debriefing of patients and their long-term follow-up are discussed fully.

If there are possible risks to human subjects, risks vs. benefits are assessed. If the committee has any reservations regarding the use or protection of human subjects, it will recommend that the provisions for the protection of human subjects be deemed inadequate (i.e., "Further Action Required"), and it will specify the changes needed for protection to be considered adequate. In addition, the applicants should demonstrate that the human subject issues have been reviewed favorably by their Institutional Review Board.

This section will be concluded with one of the following:

- Human Subjects — Protection Adequate
- Human Subjects — Exemption Claimed and Approved
- Human Subjects — Further Action Required

E. CRITIQUE

This is the most important section of the review and should embody the review committee's evaluation of the scientific and technical merit of the proposal. The critique should reflect and provide the rationale for the overall recommendation. It should present in detail both the strengths and weaknesses of the proposed research.

1. *Progress (for renewal and supplemental applications):* This section is to be written only for completing renewal and supplemental applications. The progress made in the previous grant period(s) and the relationship of the work originally approved to the work presently proposed are evaluated. Is the project proceeding in a logical manner? Are reasonable conclusions being drawn? If the aims and/or approach have changed, are they justified and appropriate? If difficulties or unexpected results have been encountered, has the applicant made appropriate changes in the renewal or supplemental application?
2. *Adequacy and relevance of background:* Is there sufficient background material to understand the reference point from which the applicant develops his/her rationale for the proposed research? Is the reference material relevant, and does it justify the direction taken by the investigator?
3. *Adequacy and feasibility of approach:* In view of the overall objectives, is the proposed approach reasonable? Is it likely to be carried out and completed without any major complications or hindrances? Does the applicant recognize and present solutions for potential problems which might arise?
4. *Hypotheses and study design:* Are the hypotheses appropriate and valid in view of the background material presented? Are the aims logical? Is the study likely to answer the questions posed? Is the study design appropriate to test the hypotheses presented? Are the protocols presented in sufficient detail? Are expected outcomes discussed? Are alternate hypotheses or protocols discussed in the event that unexpected results are obtained?
5. *Sample population and controls:* Is the proposed study population appropriate? Are the sampling techniques acceptable? Is there an appropriate control or comparison group?
6. *Data collection and analysis:* Are there any special problems related to either the collection or the analysis of data? Are standard instruments being used or are new ones being developed? Is the general approach to data reduction and analysis presented clearly? Is the analysis appropriate to both the quality and quantity of data which have been or will be collected?
7. *Significance of study:* The significance of the proposed work, relative to existing knowledge and theory, should be discussed. What is the potential importance of the proposed work to the field? The review committee discusses any novel ideas or contributions which the project offers.

F. PERSONNEL

1. *Principal investigator and senior staff:* The committee conducts a discussion of the background and competence of the project staff, with particular emphasis on the principal investigator and other key personnel. Specific positive or negative comments about any relevant special expertise (or lack of it) which enhances (or detracts from) the likelihood of the successful conduct of the project are analyzed. Is the principal investigator qualified by background, training, and experience to direct the proposed research? Is the principal investigator allocating a sufficient proportion of his time to the project?
2. *Support Staff:* Is the project staffed appropriately in number and in skills required?

G. RESOURCES AND ENVIRONMENT

The facilities and resources available for the project, including laboratories, clinical resources, computers, office space, animal quarters, special equipment, and so forth, are assessed briefly. The appropriateness of the environment (location, interdepartmental co-operation, availability of needed equipment or subjects, and the availability of expert advice, consultation, or guidance) where relevant to the application is also addressed.

H. BUDGET

1. *Appropriateness and justification:* Is the budget appropriate in terms of the aims of the project, the resources required, and the proposed work to be done? Are all items justified? Major and expensive items of equipment, consultants, costly supplies, and foreign travel should be particularly well justified. Projects funded will normally be supported for a period of up to 3 years, except in those cases where a longer commitment (a period of no longer than 5 years) may be necessary because of the nature of the research design. The review committee must justify a recommendation of support for a period exceeding 3 years.
1. *Recommended changes (if applicable):* The reviewers itemize by budget category and provide specific reasons for any recommended reductions (or in rare cases, increases) in time or amount of funding.

I. RECOMMENDATION AND RESUME

Finally, the IRG presents a concise summary of the application and states its recommendation: APPROVAL, DEFERRAL, or DISAPPROVAL. It enumerates the reasons for its recommendation and outlines any recommendations for changes in time or amount.

1. *Approval:* This is the appropriate recommendation when an application meets the standards for scientific merit and the review criteria discussed above. Rarely, approval may be recommended on the condition that certain minor changes (e.g., small alterations in methodology, deletion of some experimental approaches, additional safeguards for human subjects) be made. In such cases, the committee will present a clear statement of the recommended changes it wishes to be communicated to the investigators.
2. *Disapproval:* This is the appropriate recommendation when an application does not meet the review committee's standards for scientific merit and the criteria discussed above. Disapproval also may be recommended when the committee feels that gravely hazardous or unethical procedures are proposed.
3. *Deferral:* This is the appropriate recommendation when an application does not provide enough information to allow an evaluation of scientific merit. Such information may be obtained by correspondence or by a project site visit.

REFERENCES — SECTION V

1. Application for Public Health Service Grant (PHS 398), U.S. Department of Health and Human Services, 1986.
2. **Pandey, C.,** *Grants Magazine,* 11, 28, 1988.
3. **Goldstein, M.,** *Grants Magazine,* 10, 13, 1987.
4. **Murry, J. P. and Biles, B. R.,** *Tips for Proposal Writers,* Graduate Services & Publications, Manhattan, KS, 1980.
5. **Jagger, J.,** How to write a research proposal, *Grants Magazine,* 3, 216, 1980.
6. **Abeles, F. B.,** How to prepare an effective scientific research proposal, *Grantsmanship Center News,* 1982.

7. **Eaves, G. N.,** The research grant application: an exercise in scientific writing, *Grants Magazine*, 1, 61, 1978.
8. National Institute on Drug Abuse, Guide for Reviewer's Preliminary Comments for Research Grant Applications, DABR Review Committee, 1987.

VI. INCREASING RESEARCH PROPOSAL FUNDING COMPETITIVENESS

The purpose of this section is to describe a procedure whereby proposals can be reviewed and the resulting priority score significantly improved, resulting in a higher probability of funding. The first phase, the internal review, involves faculty members who are experienced in reviewing proposals or those who are funded by the agency in question. These preliminary reviewers are given a 5 to 10 day time period to review the proposal. These scientists will provide invaluable criticism as it relates to the clarity of the specific aims, resources and environment, budget, justification, and references. The involvement of other individuals such as scientific editor, statistician, environmental health and safety officer, and the department/division chairperson is also helpful. The resulting comments and suggestions from the initial reviewer are then given to the principal investigator, who makes the appropriate changes and prepares the proposal for the next phase, the external review. The external reviewers utilize the format of the regular NIH grants for communicating their written comments and suggestions and assigning a priority score. In addition, the proposal is reviewed, where applicable, by oversight committees (e.g., radiation) and the agency official who provides the presubmission review. The principal investigator corrects the final draft of the proposal to reflect the resulting suggestions.

This section therefore describes the procedures and guidelines used for the preparation of grant proposals which have been very effective in increasing the competitiveness of research proposals. A successful proposal depends on the development of an idea based on a thorough literature search, utilization of the appropriate personnel, and the use of methodology with sufficient detail and clarity. The development of an idea into a competitive proposal requires the allocation of adequate time for both writing and the presubmission review by advisers, administrators, scientists, a scientific editor, a statistician, and perhaps an adminstrator of the sponsoring agency. This type of quality control procedure ensures a well-written proposal that should be free of errors of omission and equipment duplication among an active faculty group. The procedures and guidelines for the preparation of grant proposals are supported by a flow diagram which depicts the interrelationship of the 12 steps or tasks of the process. The procedures and guidelines provide both directions and explanations, which are keyed by numbers to the 12 tasks in the process.

A. PROCEDURES AND GUIDELINES FOR THE PREPARATION OF GRANT PROPOSALS

The procedures described in this section are keyed by numbers to the 12 steps stated in the flow diagram (Figure I).

- Step 1. **DEVELOP THE IDEA.** Developing an idea for a proposal is the most important step. It may be helpful for you to discuss the idea with the divisional/departmental chairperson and others to determine if it is viable. This decision will involve consideration of the appropriateness of the proposal as it relates to the existing facilities and, of course, the qualifications of the investigator. If the preliminary assessment is positive, proceed to develop a competitive proposal.

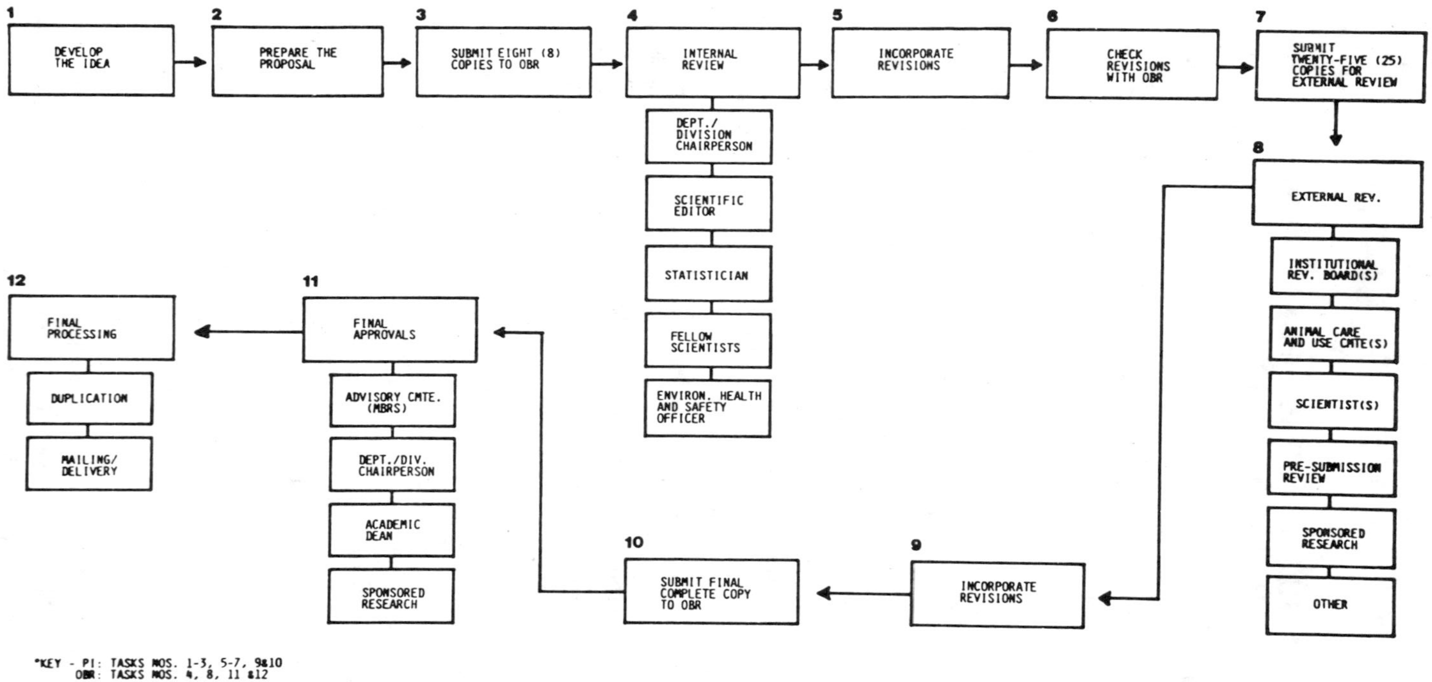

FIGURE I. Florida A&M University proposal flow diagram.

- Step 2. **PREPARE THE PROPOSAL.** The sections to be included in most proposals are shown in and the order of the sections depends on the guidelines and degree of formality required by the sponsor. The sponsor may require the completion of certain forms (i.e., U.S. Public Health Service forms) which may be peculiar.

  Experience has shown that investigators often fail to read and digest the directions provided by the sponsor. Important sections or forms may be missing or incomplete. An incomplete application cannot be processed through either the internal or external review. A very important consideration is the level of clarity and precision of the methodology, the budget, and the budget justification. The methodology must have sufficient detail as to demonstrate both an understanding of the pitfalls and the investigator's potential for successfully executing the experiments. The literature cited must be both current and thorough.

  The major line items (e.g., personnel, equipment) reflected in the budget must be shown to be integrally involved in the methodology. Moreover, the budget justification must reflect the strong dependence of the project on the requested line items. In summary, details sell.
- Steps 3 and 4. **INTERNAL REVIEW.** These steps ensure that the product sent to external reviewers is the best that the investigators can provide. Fellow scientists who were not involved in the development of the proposal critically evaluate and make suggestions on the proposal. While the area of research addressed in the proposal may not be the reviewer's specific area of research, he/she will be able to make useful observations on the focus, clarity, and completeness of the proposal.

  The scientific editor and scientist both represent a new set of eyes to assist the investigator in identifying and correcting ambiguity, syntactical errors, and omissions. The environmental health and safety officer will ensure the inclusion of adequate plans for carrying out hazardous procedures, the entry of line items in the budget for safety devices, etc. The departmental/divisional chairperson will be able to provide input regarding the issues of science and the appropriateness of the proposal for the academic unit and the level of release time requested. The statistician will ensure the appropriateness of both the experimental design and the statistical tests to be applied to the data. It should be noted that both the scientific editor and the statistician can (and should) become involved in the early stages of proposal development. This task requires a minimum of 5 to 10 days.

  Please note that the recommendations of the reviewers are suggestions for improvement. These suggestions may be written on the proposal itself or on the evaluation form. The evaluation form prepared by the primary reviewer should summarize all recommendations. If the investigator feels that particular recommendation(s) are not warranted, then he/she should indicate that to the primary reviewer or to the Office of Biomedical Research (OBR).
- Steps 5 and 6. Incorporate revisions based on internal reviewers' comments and check revisions with the OBR.
- Steps 7 and 8. **EXTERNAL REVIEW.** A clean, revised copy of the first draft should be forwarded to external reviewers as a means of ensuring that a good proposal is made more competitive before it is submitted to the sponsor. As with the internal review, recommendations represent suggestions. If the recommendations are not incorporated, the primary reviewer and the program director should be informed in writing of the reasons why the recommendations were not incorporated in the proposal. This task requires a minimum of 30 days.

  Research proposals involving human subjects must be reviewed for approval/disapproval by the Institutional Review Board (IRB) regardless of the location (on/off campus) for the research activity. If the nature of the involvement of human subjects changes as a result of external review by a scientist, it is the responsibility of the investigator to resubmit the proposal and forms to the IRB(s) for approval or disapproval. Similar requirements pertain to the local and off-campus Animal Care and Use Committee.

The peer review of the proposal by a scientist(s) is intended to result in significant improvement of the science aspect of the proposal and, of course, its potential for funding. The scientist(s) should be active in the area of research represented in the proposal and/or experienced in reviewing proposals for the potential sponsor. This person must not be a person whose objectivity can be questioned based on his/her professional or personal relationship with the investigator. The investigator is encouraged to recommend a reviewer(s) to the OBR by submitting a copy of the person's curriculum vitae. Subsequently, the potential reviewer is contacted by the OBR and a formal invitation extended.

The reviewer is expected to prepare a written critique, which will include recommendations and a priority score. A sample of instructions for Reviewer's Comments on Research Projects is presented in the next section. The priority score will, of course, indicate the potential for success of the proposal. Since some program/project applications have a maximum number of subprojects, the priority score may be used to determine the ranking of the various research components. In any case, only competitive proposals will be submitted to the sponsor for funding considerations.

Presubmission review entails the evaluation of the proposal by an administrator (e.g., health scientist administrator) in the sponsor's agency. This review should identify any concerns of the sponsor such as program/initiative relevance, budget, etc. The review of the proposal by Sponsored Research allows an early opportunity for both fiscal and institutional concerns to be identified and addressed. This task should also expedite the final processing, such as obtaining required signatures from institution officials.

- Step 9. Incorporate revisions recommended by external reviewers.
- Step 10. **SUBMIT FINAL COMPLETE PROPOSAL TO OBR.** A clean draft with appendices must be submitted to OBR with adequate lead time before the sponsor's deadline. If one is considering an appendix, keep in mind that, unlike the proposal itself, appendices will not be scrutinized by the reviewer. Thus, if the information is important, state it in the proposal.
- Steps 11 and 12. **FINAL APPROVALS AND PROCESSING.** The approval of the advisory committee, department/division chairperson, and academic dean must be obtained prior to the presentation of the proposal to Sponsored Research to secure institutional sign-off. As travel schedules of the principles cannot be predicted, 5 to 10 days are allowed for this process. The proposal is then duplicated, bound, and submitted to the sponsor. Copies are distributed to the investigator(s), departmental/divisional chairpersons, deans, and Sponsored Research after submission to the sponsor.

The final days of proposal preparation are usually frantic. If OBR has to deal with late major changes, etc., then it is likely that the final tasks will be adversely affected, with the result being overnight or same day delivery (via air freight and courier) to the sponsor. Both methods of delivery are expensive and exert a deleterious effect on the OBR budget. Moreover, a mad rush will result in an increase in the potential for a serious error.

B. RESULTS AND DISCUSSION

Prior to the initiation of a rigorous review procedure, the comments of review committees showed a clear need to improve the clarity and organization of the narrative and to increase the level of detail found in the experimental design, statistical analysis, and budget. The challenge of properly focusing the project and developing research ideas is perennial, especially for recent graduates. In 1980, prior to the implementation of this procedure, a program project proposal entitled "Biochemical and Pharmacological Investigation" earned a priority score of 250. A version of the procedures reported herein, which entailed a review only by fellow faculty members, was applied to a supplement of the above-mentioned proposal in 1981. This proposal received a priority score of 219. In 1983 both the internal and external review procedures were employed on the renewal application, "Biochemical and Pharmacological Investigations", which subsequently received a priority score of 198. In that same year a proposal was developed to host a national scientific meeting, the Minority Biomedical Research Support Symposium, and it received a priority score of 198. A 1984 supplement to the "Biochemical and Pharmacological Investigations" application received a priority score of 206. In 1985 applications for two proposals, "Florida A&M University Research Center" and "Chronopharmacological and Chronotherapeutic Investigations", were reviewed and subsequently received priority scores of 165 and 126, respectively. Thus, it appears that improvement in the priority scores resulted from this review procedure. In 1988/1989 the College secured over $7 million for research activities. Part of this success in attracting research dollars can be attributed to this vigorous proposal review process adopted by the College.

C. INSTRUCTIONS FOR REVIEWER'S COMMENTS ON RESEARCH PROJECTS*

Brief Description of the Project

Use or paraphrase the investigator's abstract.

____

____

____

____

____

Brief Description of Student Participation

List the number and types of students (graduate, undergraduate) who will participate. Describe what the students are expected to do in the proposed research.

____

____

____

____

____

* Actual space allocated for answers is substantially greater than shown.

Human Subjects

If human subjects are to be used in the project, assess the appropriateness of the subject population. Explain any physical, psychological, social, or legal risks to individuals who are participating as research subjects. Describe the procedures to minimize such risks, and discuss whether the risks are reasonable in relation to the anticipated benefits to the subjects and the importance of the knowledge that may reasonably be expected to result. Assess the adequacy of the consent procedure.

If exemptions are claimed, evaluate the appropriateness of the claimed exemptions.

Animal Welfare

If vertebrate animals are to be used in the project, discuss the appropriateness of the choice of species and number involved and the justification for their use. Assess whether the animals will receive proper care and maintenance and will not suffer unnecessary discomfort, pain, or injury. Special attention should be provided when the proposed research involves dogs, cats, nonhuman primates, or large number of animals.

Hazardous Materials and Procedures

Describe any potentially hazardous materials and procedures and whether the protection to be provided will be adequate.

Critique of Project (Regular Investigators)

This section is not to be a repeat of the project description. Evaluative statements regarding each of the areas below are requested.

A. The Problem and Its Background.

Is the hypothesis clear? Is the hypothesis biomedically relevant? Is the literature review adequate? Are the citations current?

B. Research Design and Resources: Comment on the appropriateness and adequacy of the proposed design, methods, study groups, controls, data collection and statistical analysis, equipment, and facilities. Can the aims be accomplished and the hypothesis tested conclusively, and if not, then why not?

C. Renewal Applications for Continuing Investigators. Evaluate the past progress of continuing investigators.

1. Did the investigator publish refereed articles? How many?

2. Did the investigator submit an application(s) (see other support form) for non-MBRS support? List the proposal title(s), the agency(ies), and the date of submission.

3. Student training.
   a. Have undergraduate students been motivated to pursue careers in biomedical research and/or the health professions? If not, why not?

   b. Has the investigator had experience with and/or been successful in working with MBRS or MBRS eligible students?

c. Have graduate students made presentations at national meetings other than MBRS? If not, why not?

____

____

____

Critique of Associate Investigator (Omit for Regular Investigator). See definition of Associate Investigator.

A. List the assocate investigator's active, peer-reviewed biomedical research support. Is it related to the proposed research? If not, state the difference(s).

____

____

____

B. Has the associate investigator had experience with and/or been successful in working with MBRS or MBRS eligible students?

____

____

____

Critique of Student Research

A. Do not repeat the description of student participation. Evaluate what the students are expected to do in the proposed research. Are the methods and techniques being proposed suitable for students?

____

____

____

B. Evaluate the quality of the environment, including the student's opportunities for contact with a variety of scientists and the availability of the necessary resources (space, equipment, biological models, research support, etc.).

____

____

____

____

C. Evaluate the investigator(s) commitment and competence to guide and motivate students. Will students be involved in research in ways that will promote development of careers in biomedical research and/or the health professions, and if not, why not?

Budget

Evaluate justifications for requested budget items.

For associate investigators the budget should reflect allowable costs for this category of investigator. (See Definition of Associate Investigator.)

Budget Modifications

Describe any recommended budget reduction, deletion, or addition, and reasons for the recommended changes.

Administrative Note: Identify any apparent scientific or budgetary overlap with active or pending support.

Summary and Recommendations

Summarize the strengths and weaknesses of the project. Provide the key reasons for your recommendation of approval, disapproval, or deferral.

Approval ______ Disapproval ______ Deferral ______

Priority Scores

1.0 — 1.4	Highest quality, meritorious
1.5 — 1.9	High quality
2.0 — 2.4	Better than average
2.5 — 2.9	Average
3.0 — 3.4	Slightly below average
3.5 — 3.9	Below average
4.0 — 5.0	Poor quality

Scientific merit ______

MBRS Program Relevance ______
(Consider the biomedical relevance and the student involvement.)

Note: Associate Investigators are rated only for MBRS program relevance.

A. Strengths

____

____

____

____

____

____

B. Weaknesses

____

____

____

____

____

____

VII. SAMPLES OF FUNDED AND UNFUNDED RESEARCH PROJECTS

Samples of these projects are included to assist the new researcher in preparing a research proposal.

Funded projects:

A. The Antibacterial Activity of *Brevibacterium linens*
-Funded by National Institutes of Health

B. Reducing Medication Administration Errors in Family Care Homes Through a Program of On-Site Education and Technical Assistance
-Funded by [illegible]

Approved but unfunded project:

C. Smoking Prevention in Black Children
-Submitted to [illegible]
1. Letter regarding decision on the research proposal
2. Reviewer comments and critique

A. EXAMPLE OF AN APPROVED AND FUNDED PROJECT: THE ANTIBACTERIAL ACTIVITY OF *BREVIBACTERIUM LINENS*

FORM APPROVED
O.M.B. NO. 68-R0249

DEPARTMENT OF HEALTH AND HUMAN SERVICES PUBLIC HEALTH SERVICE GRANT APPLICATION FOLLOW INSTRUCTIONS CAREFULLY	LEAVE BLANK TYPE / ACTIVITY / NUMBER REVIEW GROUP / FORMERLY COUNCIL/BOARD (Month, year) / DATE RECEIVED
1. TITLE OF APPLICATION (Do not exceed 56 typewriter spaces) THE ANTIBACTERIAL ACTIVITY OF BREVIBACTERIUM LINENS	
2. RESPONSE TO SPECIFIC PROGRAM ANNOUNCEMENT ___ NO ___ YES (If "YES," state RFA number and/or announcement title)	
3. PRINCIPAL INVESTIGATOR/PROGRAM DIRECTOR	
3a. NAME (Last, first, middle) Lewis, Bryan Aaron	3b. SOCIAL SECURITY NUMBER [redacted]
3c. MAILING ADDRESS (Street, city, state, zip code) School of Pharmacy Florida A&M University Tallahassee, Florida 32307	3d. POSITION TITLE Assistant Professor 3e. DEPARTMENT, SERVICE, LABORATORY OR EQUIVALENT School of Pharmacy
3f. TELEPHONE (Area code, number and extension) 904/599-3343	3g. MAJOR SUBDIVISION
4. HUMAN SUBJECTS, DERIVED MATERIALS OR DATA INVOLVED [X] NO [ ] YES (If "YES," form HHS 596 required)	5. RECOMBINANT DNA RESEARCH SUBJECT TO NIH GUIDELINES [X] NO [ ] YES
6. DATES OF ENTIRE PROPOSED PROJECT PERIOD (This application) From: July 1, 1982 Through: June 30, 1985	7. TOTAL DIRECT COSTS REQUESTED FOR PROJECT PERIOD (from page 5) $ 81,395 8. DIRECT COSTS REQUESTED FOR FIRST 12-MONTH BUDGET PERIOD (from page 4) $26,409
9. PERFORMANCE SITES (Organizations and addresses) Florida A&M University School of Pharmacy Post Office Box 367 Tallahassee, Florida 32307	10. INVENTIONS (Competing continuation application only) Were any inventions conceived or reduced to practice during the course of the project? [ ] NO [ ] YES - Previously reported [ ] YES - Not previously reported 11. APPLICANT ORGANIZATION (Name, address, and congressional district) Florida A&M University IRS #[redacted] Post Office Box 367 Tallahassee, Florida 32307 2nd Congressional District
12. ORGANIZATIONAL COMPONENT TO RECEIVE CREDIT FOR INSTITUTIONAL GRANT (See instructions) Code 0 7 Description	13. ENTITY IDENTIFICATION NUMBER 14. TYPE OF ORGANIZATION (See instructions) [ ] Private Nonprofit [X] Public (Specify Federal, State, Local):
15. OFFICIAL IN BUSINESS OFFICE TO BE NOTIFIED IF AN AWARD IS MADE (Name, title, address and telephone number.) Mr. Frank C. Olds Sponsored Research Florida A&M University Tallahassee, Florida 32307	16. OFFICIAL SIGNING FOR APPLICANT ORGANIZATION (Name, title, address and telephone number) Dr. Walter L. Smith, President Florida A&M University Tallahassee, Florida 32307 904/599-3225
17. PRINCIPAL INVESTIGATOR/PROGRAM DIRECTOR ASSURANCE: I agree to accept responsibility for the scientific conduct of the project and to provide the required progress reports if a grant is awarded as a result of this application.	SIGNATURE OF PERSON NAMED IN 3a (In ink. "Per" signature not acceptable) Bryan A Lewis DATE 10/29/81
18. CERTIFICATION AND ACCEPTANCE: I certify that the statements herein are true and complete to the best of my knowledge, and accept the obligation to comply with Public Health Service terms and conditions if a grant is awarded as the result of this application. A willfully false certification is a criminal offense. (U.S. Code, Title 18, Section 1001.)	SIGNATURE OF PERSON NAMED IN 16 (In ink. "Per" signature not acceptable) Walter L. Smith DATE 10/29/81

PHS-398
Rev 5/80

DEPARTMENT OF HEALTH AND HUMAN SERVICES
PUBLIC HEALTH SERVICE

ABSTRACT OF RESEARCH PLAN

LEAVE BLANK

PROJECT NUMBER

NAME AND ADDRESS OF APPLICANT ORGANIZATION *(Same as Item 11, page 1)*

Florida Agricultural and Mechanical University
Tallahassee, Florida 32307

TITLE OF APPLICATION *(Same as Item 1, page 1)*

Name, Title and Department of all professional personnel engaged on project, beginning with Principal Investigator/Program Director

Bryan A. Lewis, Ph.D. Assistant Professor School of Pharmacy

Milton Cofield, Ph.D. Research Chemist, Kodak Research Laboratories, Consultant

ABSTRACT OF RESEARCH PLAN: Concisely describe the application's specific aims, methodology and long-term objectives, making reference to the scientific disciplines involved and the health-relatedness of the project. The abstract should be self-contained so that it can serve as a succinct and accurate description of the application when separated from it. DO NOT EXCEED THE SPACE PROVIDED.

With the increasing emergence of antibiotic-resistant microbial strains, it is imperative that the search for new and effective antibiotics continue. Thus, the overall objective of this project is to investigate antibiotic production by *Brevibacterium linens*. The results could possibly yield information regarding a new and useful antibiotic. *Brevibacterium linens* is a bacteria that compose part of the surface flora of aged surface-ripened cheese such as Limburger and Liederkrang and has been shown to produce a substance(s) which is inhibitory to a wide range of bacteria, molds, and yeasts. A few properties of the agent are known but the identity of the agent(s) remain unknown. This investigation is designed to examine specifically, the antibacterial activity of *B linens* and attempt to isolate, purify and characterize the agent(s) responsible.

Various strains of *Brevibacterium linens* will be grown in several production media and culture filtrates tested for antibacterial activity by challenging the filtrates against clinically significant indicator bacteria using agar-diffusion and tube-dilution techniques. After screening for antibacterial activity, attempts will be made to isolate and purify the agent using chromatographic and electrophoretic techniques. The physio-chemical properties of the agent will be analyzed by chromatographic and spectroscopic methods (IR, UV, NMR, elemental analysis) and it is anticipated that results will elucidate the molecular structure of the agent.

The research is significant in that the antibacterial agent produced by *Brevibacterium linens* could be a potential chemotherapeutic agent and supplement present antimicrobial agents. In addition to a potential medical value, it may have use in a non-medical capacity (agent for food preservation, etc.)

LABORATORY ANIMALS INVOLVED. Identify by common names. If none, state "none"

PHS-398
Rev. 5/80

PRINCIPAL INVESTIGATOR PROGRAM DIRECTOR: Lewis, Bryan A.

TABLE OF CONTENTS

Number pages consecutively at the bottom throughout the application. Do not use suffixes such as 5a, 5b. Type the name of the Principal Investigator/Program Director at the top of each printed page and each continuation page.

SECTION 1. PAGE NUMBERS

SECTION 2.

SECTION 3. Appendix *(Six sets) (No page numbering necessary for Appendix)*

Number of publications: ______ Number of manuscripts: ______
Other items (list):

Application Receipt Record, form PHS 3830
Form HHS 596 if Item 4, page 1, is checked

PHS-398
Rev. 5/80

PRINCIPAL INVESTIGATOR/PROGRAM DIRECTOR:

DETAILED BUDGET FOR FIRST 12 MONTH BUDGET PERIOD
DIRECT COSTS ONLY

FROM	THROUGH
July 1, 1982	June 30, 1983

DOLLAR AMOUNT REQUESTED *(Omit cents)*

PERSONNEL *(Applicant organization only) (See instructions)*

NAME	TITLE OF POSITION	TIME/EFFORT %	Hours per Week	SALARY	FRINGE BENEFITS	TOTALS
Bryan A. Lewis, Ph.D.	Principal Investigator	50%	A/Y	[redacted]	[redacted]	[redacted]
		50%	(Summer)	[redacted]	[redacted]	[redacted]
			SUBTOTALS →	[redacted]	[redacted]	[redacted]

Category	Description	TOTALS
CONSULTANT COSTS *(See instructions)*	One(1) Consultant @ $150/day X 2 Days (#300) plus Travel	600
EQUIPMENT *(Itemize)*	Refrigerator - $800 pH Meter - $600	1400
SUPPLIES *(Itemize by category)*	Chemicals & Media - $2000 Bacterial Cultures - $300 Glassware - $1200	3500
TRAVEL — DOMESTIC	One (1) Faculty X One (1) Trip - $500/trip	500
TRAVEL — FOREIGN		
PATIENT CARE COSTS — INPATIENT		
PATIENT CARE COSTS — OUTPATIENT		
ALTERATIONS AND RENOVATIONS *(Itemize by category)*		
CONTRACTUAL OR THIRD PARTY COSTS *(See instructions)*		
OTHER EXPENSES *(Itemize by category)*	Publication Costs - $250	250
TOTAL DIRECT COSTS *(Also enter on page 1, item 8)* →		$ 26409

PHS-398
Rev. 5/80

PRINCIPAL INVESTIGATOR PROGRAM DIRECTOR: Lewis, Bryan A.

BUDGET ESTIMATES FOR ALL YEARS OF SUPPORT REQUESTED
DIRECT COSTS ONLY

BUDGET CATEGORY TOTALS		1st BUDGET PERIOD (*from page 4*)	ADDITIONAL YEARS SUPPORT REQUESTED 2nd	3rd	4th	5th
PERSONNEL (*Salary and fringe benefits.*) (*Applicant organization only*)		[redacted]	[redacted]	[redacted]		
CONSULTANT COSTS		600	600	600		
EQUIPMENT		1400	-0-	-0-		
SUPPLIES		3500	3500	3500		
TRAVEL	DOMESTIC	500	500	500		
	FOREIGN	-0-	-0-	-0-		
PATIENT CARE COSTS	INPATIENT	-0-	-0-	-0-		
	OUTPATIENT	-0-	-0-	-0-		
ALTERATIONS AND RENOVATIONS		-0-	-0-	-0-		
CONTRACTUAL OR THIRD PARTY COSTS		-0-	-0-	-0-		
OTHER EXPENSES		250	250	250		
TOTAL DIRECT COSTS		$26409	$26622	$28364		

TOTAL FOR ENTIRE PROPOSED PROJECT PERIOD (*Also enter on page 1, item 7*) → $81,395

JUSTIFICATION (*Use continuation pages if necessary*): Briefly describe the specific functions of the personnel and consultants. For all years, justify any costs for which the need may not be obvious, such as equipment, foreign travel, alterations and renovations, and contractual or third party costs. For future years, justify any significant increases in any category. In addition, for **COMPETING CONTINUATION** applications, justify any significant increases over current level of support. If a recurring annual increase in personnel costs is anticipated, give percentage.

BUDGET JUSTIFICATION:

Refrigerator: There is a refrigerator presently in the laboratory that is being shared with several other faculty members and therefore utilized to full capacity. The refrigerator being requested will be used to house only research material and microorganisms used in this investigation.

pH Meter: There are pH meters in the school which are maximally used and are not easily accessible for my lab. A pH meter to be used for microbiological research only and readily accessible will diminish time-consuming searches.

Consultant: The consultant expertise will be necessary and cruical for disseminating information regarding the isolation, purification and characterization of the agent. He is a chemist and his advise is invaluable.

PHS-398
Rev. 5/80

PRINCIPAL INVESTIGATOR/PROGRAM DIRECTOR: Lewis, Bryan A.

BIOGRAPHICAL SKETCH

Give the following information for key professional personnel listed on page 2, beginning with the Principal Investigator/Program Director. Photocopy this page for each person.

NAME	TITLE	BIRTHDATE (Mo., Day, Yr.)
Bryan A. Lewis	Assistant Professor	05/23/51

EDUCATION (*Begin with baccalaureate training and include postdoctoral*)

INSTITUTION AND LOCATION	DEGREE	YEAR CONFERRED	FIELD OF STUDY
Southern University, Baton Rouge, LA	B.S.	1973	Microbiology
McNeese State University, Lake Charles, LA	M.S.	1975	Microbiology
University of Southern Mississippi, Hattiesburg, MS	Ph.D.	1980	Microbiology

RESEARCH AND/OR PROFESSIONAL EXPERIENCE: Concluding with present position, list in chronological order previous employment, experience, and honors. Include present membership on any Federal Government Public Advisory Committee. List, in chronological order, the titles and complete references to recent representative publications, especially those most pertinent to this application. Do not exceed 2 pages.

1971 - 1972: NIH Undergraduate Trainee, Southern University

1973 - 1975: Graduate Assistant, McNeese State University

1975 - 1977: Graduate Assistant, University of Southern Louisiana, Lafayette

1979 - Present: Assistant Professor of Pharmaceutic Science

HONORS:

1977 - 1979: Predoctoral Fellowship, University of Southern Mississippi, Hattiesburg, Mississippi

1979 - 1980: Faculty Incentive Award, School of Pharmacy, Tallahassee, Florida

MAJOR RESEARCH INTEREST:

Antibiotic Production by Microorganisms, Principan Investigator

PUBLICATIONS:

Lewis, B. A., 1980, "A Study of the Antifungal Agent(s) Produced by Four Strains of "*Brevibacterium*". (Ph.D. Dissertation, Department of Microbiology, University of Southern Mississippi, Hattiesburg, Mississippi).

Lewis, B. A., 1981, A Dual Culture Method for Assaying the Antifungal Activity of *Brevibacterium linens* (Submitted to "Experientia").

Lewis, B. A., 1981, Inhibition of Several Food Spoilage Fungi by an Antifungal Agent(s) Produced by *Brevibacterium linens* (Accepted for Publication).

Lewis, B. A., 1981, Correlation with Antifungal Activity and Thiol Production by *Brevibacterium linens*. (Submitted to Antonie Van Leeuwenhoek).

Jackson, J. and B. A. Lewis, 1981, A Simplified Growth Medium for *Brevibacterium linens* (Manuscript in Preparation).

Lewis, B. A. and B. F. Fleet, 1981, Acyclovir - A New Antiherpetic Drug (Submitted to Florida Journal of Pharmacy).

PRINCIPAL INVESTIGATOR/PROGRAM DIRECTOR: Lewis, Bryan A.

OTHER SUPPORT

(USE CONTINUATION PAGES IF NECESSARY)

For each of the professionals named on page 2, list, in three separate groups: (1) active support; (2) applications pending review and/or funding; (3) applications planned or being prepared for submission. Include all Federal, non-Federal, and institutional grant and contract support. If none, state "NONE." For each item give the source of support, identifying number, project title, name of principal investigator/program director, time or percent of effort on the project by professional named, annual direct costs, and entire period of support. (If part of a larger project, provide the titles of both the parent grant and the subproject and give the annual direct costs for each.) Briefly describe the contents of each item listed. If any of these overlap, duplicate, or are being replaced or supplemented by the present application, justify and delineate the nature and extent of the scientific and budgetary overlaps or boundaries.

PRINCIPAL INVESTIGATOR/PROGRAM DIRECTOR:
(1) ACTIVE SUPPORT:

NONE

PRINCIPAL INVESTIGATOR/PROGRAM DIRECTOR: Lewis, Bryan A.

RESOURCES AND ENVIRONMENT

FACILITIES: Mark the facilities to be used and briefly indicate their capacities, pertinent capabilities, relative proximity and extent of availability to the project. Use "other" to describe facilities at other performance sites listed in Item 9, page 1, and at sites for field studies. Using continuation pages if necessary, include a description of the nature of any collaboration with other organizations and provide further information in the RESEARCH PLAN.

[X] Laboratory: The primary laboratories used in the study are Rooms 022 and 022A of the Dyson Pharmacy Building. Together, these labs house major equipment such as the Autoclave, Centrifuge, Fermentor, Shakers and other necessary supplies necessary for microbiological research. Room 022A is the designated room for microbiology research and it also contains glassware, media, incubator, and other microbiological supplies.

[ ] Clinical:

[ ] Animal:

[ ] Computer:

[ ] Office:

[ ] Other (__________):

MAJOR EQUIPMENT: List the most important equipment items already available for this project, noting the location, and pertinent capabilities of each.

Autoclave
Incubator Shaker
Fermentor
Incubators
GLC-Mass Spectrometer

These are all located in the Dyson Pharmacy Building and are fully operational.

ADDITIONAL INFORMATION: Provide any other information describing the environment for the project. Identify support services such as consultants, secretarial, machine shop, and electronics shop, and the extent to which they will be available to the project.

Dr. Milton Cofield of Kodak Research Laboratories (New York) will serve as consultant on this project. Dr. Cofield is a physical chemist and his expertise will be used for outlining methods, procedures, and techniques which might possibly serve for purfying and characterizing the antibiotic.

Name of PI/PD/Program Coordinator or Candidate (Last, first initial)	Social Security Number
Lewis, Bryan A.	[redacted]

VI. Research Plan

A. Specific Aims

Brevibacterium linens, a gram-positive nonsporeforming rod-shaped bacterium present in the flora of a variety of aged surfaced-ripened cheese has been shown to produce an agent(s) which is inhibitory to a wide range of bacteria, yeasts and molds. The assays however, used to test the antimicrobial activity of the bacterium, did not always give clear results. In addition, the substance(s) responsible for the observed inhibitory effects on the test organisms has not been identified.

Therefore, the aims of this investigations are:

To grow several strains of Brevibacterium linens in various media and screen the organisms for antibiotic production by using agas diffusion and broth dilution assay techniques.

To further develop and define the media, environmental conditions, and assay methods which would be suitable for maximum antibiotic production and clear demonstration of the inhibitory effect.

To isolate, purify, and attempt to characterize the agent(s) responsible for antibacterial activity.

B. Significance

The word antibiotic means "against life". The term was first used by Waksman in 1942 and has been defined as a microbial product of low molecular weight that has the capacity to specifically interfere with the growth of and even to destroy microorganisms when it is present in minute amounts. Most of these substances are fungal or bacterial metabolites which have no obvious role in the growth and maintenance of the cell that produces them. Their marked and selective biologic activity against microorganism as well as their low toxicity, make possible the use of some of the antibiotics to destroy microorganisms *in vivo*. Owing to these properties many antibiotics have rapidly found practical application in combating infectious diseases.

Of all the antibiotics known, only a few are produced commercially on a large scale. To be of any interest, a new antibiotic must posses features which make it distinctly superior to any already reported. Only a few antibiotics are completely non-toxic to the animal body, but the toxicity

Name of PI/PD/Program Coordinator or Candidate (Last, first initial): Lewis, Bryan A.

Social Security Number: [redacted]

of others is suficiently low to enable them to be used therapeutically; some are so toxic as to offer little promise as therapeutic agents. Antibiotics can differ in their mode of action against a number of microorganisms and also in the rapidity with which microorganisms develop resistance to them.

The importance and practical uses of antibiotics are well known. A large portion of antibiotics are used as chemotherapeutic agents and are of medical value. The rapid selection and distribution of resistant pathogenic bacteria emphasize the need for continued research for antibiotics and synthetic drugs to replace those whose effectiveness declines. Moreover, antibiotics are needed for diseases against which there is no present treatment.

Other antibiotics have found non-medical uses such as pesticides, supplements in animal feeds which stimulate the growth of animals for market, or as food preservatives which inhibit the growth of microorganisms that cause food spoilage.

At present, new antibiotics are constantly being sought. The promising ones are tested to establish their usefulness and safety in treating diseases of man, other animals and plants. This research serves a germane purpose in that the end results of the study could possibly give rise to a new and useful antibiotic.

Investigations concerning the antimicrobial agent(s) found in aged surface-ripened cheeses were undertaken by Grecz, Dack and Hedrick in 1961 and 1962. Their findings demonstrated that the source of the antimicrobial agent(s) was Brevibacterium linens and that the agent's properties were the same as those of the antimicrobial agent(s) extracted from aged Liederkranz cheese (Grecz et. al., 1962). A few properties of the agent(s) were reported but the chemical identity of the agent(s) was not ascertained (Grecz et. al., 1962).

It is not known if there are one or more agents responsible for the antimicrobial activity. The agents produced by Brevibacterium linens and responsible for antifungal activity was investigated by Lewis in 1979 (Lewis, Ph.D. Dissertation). This investigation found a correlation between thiol production and antifungal activity by the organism (Lewis, unpublished data). Although this suggests thiols as antimycotics it does, by no means, also suggest thiols as antibacterial agents. It is at present, not known if the agent responsible for antifungal activity is also responsible for antibacterial activity.

The present investigation will study specifically the agent produced by Brevibacterium linens and responsible for antibacterial activity. This is significant because the agent(s) extracted by Grecz et. al. in 1962 inhibited the growth of some medically important bacteria.

Name of PI/PD/Program Coordinator or Candidate (Last, first initial)	Social Security Number
Lewis, Bryan A.	[redacted]

The importance of this research is apparent in that it could possibly yield a valuable antibacterial antibiotic which would supplement the existing antibiotics used in a therapeutic and/or non therapeutic capacity. Because the agent(s) is consumed when eating Limburger and similar varieties of cheeses it is possible that it could be nontoxic in pure form. Of course, only investigations directed in this direction and clinical trials with the purified antibiotic could confirm this hypothesis.

C. Preliminary Studies

The antifungal activity of Brevibacterium linens has been studied by Lewis (1979). It was found in these investigations that several molds were inhibited by products of Brevibacterium linens and that there is a correlation between antifungal activity and thiol production by these strains. I observed that Brevibacterium linens ACTT 8377 also demonstrated marginal antibacterial activity against Staphylococus aureus. The extent of this inhibitory activity by this strain has not been fully explored. At present, a few strains of Brevibacterium linens have been isolated from the mixed surface microflora of Limburges cheese samples. A selective medium for Brevibacterium linens is now being developed to facilitate isolation of the organisms from the cheese samples.

D. Methods

Isolation and Characterization of the Producing Bacteria

Different strains of antibiotic-producing microorganisms may vary in their production of antibiotics. In accordance with this, several strains of Brevibacterium linens will be obtained from the American Type Culture Collection (ATCC) and others isolated from the surface of Limburger cheese samples. Taxonomic studies of the cheese isolates will be carried out according to the Manual of Microbiological Methods and tests described by Sharpe et. al (1977). The taxonomic studies will include examining the microscopic, cultural, and physiological characteristics of the isolated bacteria. Microscopic examinations include ascertaining the morphological and straining reactions of the bacteria. Cultural characteristics includes pigmentation, colonial forms (smooth or rough) when grown on media and growth characteristics in broth media (pellicle, sediment, etc.). The physiological characteristics of the organism that will be tested include the organisms utilization of carbohydrates, their salt tolarance, DNAse activity, etc. Classification of the isolates will be done according to Bergey's Manual of Determinative Bacteriology, 7th Ed. The test or indicator bacteria to be used for assaying the antibacterial activity possessed by the Brevibacterium linens cultures will be obtained from the American Type Culture Collection.

Name of PI/PD/Program Coordinator or Candidate (Last, first initial)	Social Security Number
Lewis, Bryan A.	[redacted]

Antibiotic Production Media

Microbial growth is necessary for antibiotic production therefore, it is important that suitable culture media along with the proper environmental conditions (pH, temperature aeration) are used. The environmental conditions required for growth of Brevibacterium linens are known and will be used in these studies. Different media can have a profound influence on antibiotic production and because of this several production media will be used. Both complex and defined media will be used. Production media that will be used include the following:

1. Nutrient Broth
2. Brain Heart Infusion
3. Peptone (0.5%) Yeast Extract (0.25%)
4. Tryptone Glucose Yeast Extract
5. Casamino Acids (0.5%) Yeast Extract (0.25%)
6. Casein
7. Corn Steep Liquor
8. Soybean Meal
9. Tryptic Soy Broth
10. Tryptone (0.5%) Yeast Extract (0.25%)

Preliminary Screening for Antimicrobial Activity

Each strain of Brevibacterium linens will be grown in the aforementioned production media and tested for their ability to produce an antimicrobial agent. Bacterial incubation will take place in 250 ml of the various production media and 10 ml of aliquots removed after specified time intervals (daily). The aliquotes will be centrifuged at 8,000-10,000 rpm for approximately 20 minutes to pellet the bacteria and obtain culture supernatants. These supernatants will be tested for antimicrobial activity primarily using agar-diffusion assay methods. Two agar-diffusion methods will be employed, that of the paper-disc and well-diffusion assays. The disc-agar diffusion method consists of saturating a filter paper disc with respective culture supernatant and placing the disc onto plates previously seeded (by swabbing the entire plate) with the test or indicator bacteria. The indicator bacteria will include members from the genera staphylococcus, streptococcus, Escherichia, Pseudodomonas, Proteus and Serratia. Antimicrobial activity is demonstrated by a clear zone surrounding the disc. The clear zone represents absence of bacterial growth and is regarded as index of the sensitivity of the test bacteria to the antibiotic or the potency of the antibiotic. The well diffusion method is similar to the paper disc method but its difference lies in loading aliquots of culture supernatant into wells bored into agar media previously seeded with test bacteria. The medium on which the test bacteria will be grown is Mueller-Hinton agar. There could be problems encountered by the agar diffusion assays such as the antimicrobial agent not diffusing well through the agar or a retardation of the agent by the paper disc. This disadvantage could possibly be circumvented by testing the inhibitory

Name of PI/PD/Program Coordinator or Candidate (Last, first initial)	Social Security Number
Lewis, Bryan A.	[redacted]

effect of the antimicrobial agent in a broth system. This will entail placing aliquots of heat-treated culture supernatants (to kill B. linens- - antimicrobial agent was shown to be heat-stable) into 13 X 100 ml tubes. Standardized inoculum containing the respective test bacteria will also be placed in the tubes and allowed to incubate. Growth or its absence after appropriate incubation would be an index of bacterial inhibition by antimicrobial agent. Other problems may also be encountered during the screening process but they will be met and attempts to solve them as they arise. After the preliminary screening, only the strains that possess antibacterial activity will be used for further investigations. If possible, the strain which is most "antibiotically active" will be used as the producing organism in the study.

Optimizing Antibiotic Production

The culture media and environmental conditions which are found to yield antibiotic production by the "active" strains will be used as the standard. The time required for antibiotic production in the medium and under the specific conditions will be recorded and used as a control in subsequent studies. Manipulations of the media (varying the concentrations of ingredients) and environmental conditions (changes in pH, temperature and aeration) will be undertaken to possibly optimize the production of the antimicrobial agent. These manipulations are essential if a media and conditions that would promote optimal antibiotic production are to be achieved. Antibiotic production will be measured after each manipulation of growth media and/or variation of environmental conditions again by disc agar methods. Here we seek to ascertain the time required for maximum antibiotic production with comparison to the control. If a medium and or conditions can be achieved that would promote maximum antibiotic production, then, these factors will be used to prepare the final production media.

Fermentation

The strains of Brevibacterium linens which produces the antibacterial agent will be grown on agar slants identical in composition to that of the production medium. The organisms on these slants will be the seed culture and used to inoculate a fermentor containing approximately 1000-1500 ml of production medium. The proper temperature, pH, and agitation obtained from above studies will be adjusted to the correct specification(s) and the fermentation process allowed to proceed for the time interval required for maximum antibiotic production.

Isolation and Purification

Culture supernatant will be collected from the fermentor after the appropriate incubation period by centrifugation. Centrifugation of the medium is necessary to pellet the bacteria. After the supernatant is collected, attempts will then be made to isolate and purify the agent. Extraction procedures will be performed on the culture supernatants to

Name of PI/PD/Program Coordinator or Candidate (Last, first initial): Lewis, Bryan A.

Social Security Number: [redacted]

isolate the active fractions. Organic solvents such as butanol, methanol or chloroform will be used in the extraction process. The residue obtained from the extraction will be dissolved in an organic solvent such as methanol and the solution applied to colum chromatography. Columns packed with silica gel or sephadex LH-20 are used to purify antibiotics and will be used in this study. The fractions applied to the column will be eluted with organic solvents and the active fraction(s) dried and collected. Further purification of the fractions with these columns will be done if necessary. Electrophoresis is a process which could be used to separate the fractions and also determine the charge on the agent which would allow information concerning the use of ion-exchange chromatography.

General Characterization

The physio-chemical properties of the active fraction(s) will be examined once the antibiotic is isolated and purified. Properties of the antibiotic to be ascertained include solubility in solvents, elemental analysis, and the ultraviolet and infrared absorption spectra. The techniques of ^{1}H, and ^{13}C nmr spectroscopy will be applied to help establish the antibiotics' chemical identity. Mass spectra will also be obtained to ascertain the molecular weight of the antibiotic.

Summary of Methodology

Since previous investigations have shown that antibiotic production by Brevibacterium linens is not readily demonstrated, a large portion of the research will focus on the development of appropriate growth media and conditions that will most be favorable for antibiotic biosynthesis. The testing of the antibiotics activity against a wide spectra of bacteria will also comprise an equal part. The second phase of the research consisting of the isolation, purification and characterization procedures will be done with collaboration of the consultant and other investigators who have expertise in these areas. The consultant will play a large role in this aspect of the project. He will serve to advise on methods and techniques (that are within the school's capacity to perform) which could possibly elucidate the antibiotics structure.

Also, recommendations are anticipated from him regarding outside institutions that have the capabilities (of which we do not) to perform the characterization of the antibiotic.

DEPARTMENT OF HEALTH AND HUMAN SERVICES PROTECTION OF HUMAN SUBJECTS ASSURANCE/CERTIFICATION/DECLARATION ☐ ORIGINAL ☐ FOLLOWUP ☐ REVISION	☐ GRANT ☐ CONTRACT ☐ FELLOW ☐ OTHER ☒ NEW ☐ RENEWAL ☐ CONTINUATION APPLICATION IDENTIFICATION NUMBER *(If known)*

STATEMENT OF POLICY: Safeguarding the rights and welfare of subjects at risk in activities supported under grants and contracts from DHHS is primarily the responsibility of the institution which receives or is accountable to DHHS for the funds awarded for the support of the activity. In order to provide for the adequate discharge of this institutional responsibility, it is the policy of DHHS that no activity involving human subjects to be supported by DHHS grants or contracts shall be undertaken unless the Institutional Review Board has reviewed and approved such activity, and the institution has submitted to DHHS a certification of such review and approval, in accordance with the requirements of Public Law 93-348, as implemented by Part 46 of Title 45 of the Code of Federal Regulations, as amended, (45 CFR 46). Administration of the DHHS policy and regulation is the responsibility of the Office for Protection from Research Risks, National Institutes of Health, Bethesda, MD 20205.

1. TITLE OF PROPOSAL OR ACTIVITY

THE ANTIBACTERIAL ACTIVITY OF BREVIBACTERIUM LINENS

2. PRINCIPAL INVESTIGATOR/ACTIVITY DIRECTOR/FELLOW

Bryan A. Lewis, Ph.D.

3. DECLARATION THAT HUMAN SUBJECTS EITHER WOULD OR WOULD NOT BE INVOLVED

☒ A. NO INDIVIDUALS WHO MIGHT BE CONSIDERED HUMAN SUBJECTS, INCLUDING THOSE FROM WHOM ORGANS, TISSUES, FLUIDS, OR OTHER MATERIALS WOULD BE DERIVED, OR WHO COULD BE IDENTIFIED BY PERSONAL DATA, WOULD BE INVOLVED IN THE PROPOSED ACTIVITY. (IF NO HUMAN SUBJECTS WOULD BE INVOLVED, CHECK THIS BOX AND PROCEED TO ITEM 7. PROPOSALS DETERMINED BY THE AGENCY TO INVOLVE HUMAN SUBJECTS WILL BE RETURNED.)

☐ B. HUMAN SUBJECTS WOULD BE INVOLVED IN THE PROPOSED ACTIVITY AS EITHER: ☐ NONE OF THE FOLLOWING, OR INCLUDING: ☐ MINORS, ☐ FETUSES, ☐ ABORTUSES, ☐ PREGNANT WOMEN, ☐ PRISONERS, ☐ MENTALLY RETARDED, ☐ MENTALLY DISABLED. UNDER SECTION 6. COOPERATING INSTITUTIONS, ON REVERSE OF THIS FORM, GIVE NAME OF INSTITUTION AND NAME AND ADDRESS OF OFFICIAL(S) AUTHORIZING ACCESS TO ANY SUBJECTS IN FACILITIES NOT UNDER DIRECT CONTROL OF THE APPLICANT OR OFFERING INSTITUTION.

4. DECLARATION OF ASSURANCE STATUS/CERTIFICATION OF REVIEW

☐ A. THIS INSTITUTION HAS NOT PREVIOUSLY FILED AN ASSURANCE AND ASSURANCE IMPLEMENTING PROCEDURES FOR THE PROTECTION OF HUMAN SUBJECTS WITH THE DHHS THAT APPLIES TO THIS APPLICATION OR ACTIVITY. ASSURANCE IS HEREBY GIVEN THAT THIS INSTITUTION WILL COMPLY WITH REQUIREMENTS OF *DHHS Regulation 45 CFR 46*, THAT IT HAS ESTABLISHED AN INSTITUTIONAL REVIEW BOARD FOR THE PROTECTION OF HUMAN SUBJECTS AND, WHEN REQUESTED, WILL SUBMIT TO DHHS DOCUMENTATION AND CERTIFICATION OF SUCH REVIEWS AND PROCEDURES AS MAY BE REQUIRED FOR IMPLEMENTATION OF THIS ASSURANCE FOR THE PROPOSED PROJECT OR ACTIVITY.

☐ B. THIS INSTITUTION HAS AN APPROVED GENERAL ASSURANCE (DHHS ASSURANCE NUMBER ________) OR AN ACTIVE SPECIAL ASSURANCE FOR THIS ONGOING ACTIVITY, ON FILE WITH DHHS. THE SIGNER CERTIFIES THAT ALL ACTIVITIES IN THIS APPLICATION PROPOSING TO INVOLVE HUMAN SUBJECTS HAVE BEEN REVIEWED AND APPROVED BY THIS INSTITUTION'S INSTITUTIONAL REVIEW BOARD IN A CONVENED MEETING ON THE DATE OF ________ IN ACCORDANCE WITH THE REQUIREMENTS OF THE *Code of Federal Regulations on Protection of Human Subjects (45 CFR 46)*. THIS CERTIFICATION INCLUDES, WHEN APPLICABLE, REQUIREMENTS FOR CERTIFYING FDA STATUS FOR EACH INVESTIGATIONAL NEW DRUG TO BE USED (SEE REVERSE SIDE OF THIS FORM).

THE INSTITUTIONAL REVIEW BOARD HAS DETERMINED, AND THE INSTITUTIONAL OFFICIAL SIGNING BELOW CONCURS THAT:

EITHER ☐ HUMAN SUBJECTS WILL NOT BE AT RISK; OR ☐ HUMAN SUBJECTS WILL BE AT RISK.

5. AND 6. SEE REVERSE SIDE

7. NAME AND ADDRESS OF INSTITUTION

Florida Agricultural And Mechanical University
Tallahassee, Florida 32307

8. TITLE OF INSTITUTIONAL OFFICIAL	TELEPHONE NUMBER
Dr. Walter L. Smith, President Florida A&M University	904/599-3225
SIGNATURE OF INSTITUTIONAL OFFICIAL	DATE
Walter L. Smith	10-29-81

HHS-596 (Rev. 5-80)

ENCLOSE THIS FORM WITH THE PROPOSAL OR RETURN IT TO REQUESTING AGENCY.

Name of PI/PD/Program Coordinator or Candidate (Last, first initial) | Social Security Number

Lewis, Bryan A.

E. Literature Cited

1. Breed, R. S., E. Murray and N. R. Smith. 1957. Bergey's Manual of Determinative Bacteriology 7th ed. The Williams and Wilkins Co., Baltimore.

2. Grecz, N., R. O. Wagenaar and G. M. Dack. 1957. Inhibition of Clostridium botulinum and molds in aged surfaced-ripened cheese. Appl. Microbiol., 7: 33.

3. Grecz, N., R. O. Wagenaar, and G. M. Dack. 1959. Inhibition of Clostridium botulinum by culture filtrates of Brevibacterium linens. J. Bacteriol. 78: 506-510.

4. Grecz, N., G. M. Dack and L. R. Hedrick. 1961. Antimicrobial Agent of surfaced-ripened cheese. I. Isolation and Assay. J. Food Sci. 26: 72.

5. Grecz, N., G. M. Dack and L. R. Hedrick. 1962. Antimicrobial Agent of surfaced ripened-cheese. II. Source and Properties of the Active Principle(s). J. Food Sci. 27: 335-342.

6. Korzybski, Tadeusz. 1967. Antibiotics - Origin, Nature and Properties, Pergamon Press, New York.

7. Lewis, B. A., "A Study of the Antifungal Agent(s) Produced by Four Strains of Brevibacterium. (Ph.D. Dissertation, Department of Microbiology, University of Southern Mississippi, Hattiesburg, Mississippi.

8. Sharpe, M. E., B. A. Law and B. A. Phillips. 1976. Coryneform Bacteria producing methane thiol. J. Gen. Microbiol., 94: 430-435.

9. Society of American Bacteriologist, Committee on Bacteriological Technique. 1957. Manual of Microbiological Methods. McGraw-Hill Book Co., New York.

F. Appendix

None

PRINCIPAL INVESTIGATOR/PROGRAM DIRECTOR: Lewis, Bryan A.

CHECKLIST

This is the required last page of the application.

Check the appropriate boxes and provide the information requested.

TYPE OF APPLICATION:

[X] NEW application *(This application is being submitted to the PHS for the first time.)*

[ ] COMPETING CONTINUATION of grant number: ____________________.
(This application is to extend a grant beyond its original project period.)

[ ] SUPPLEMENT to grant number: ____________________.
(This application is for additional funds during a funded project period.)

[ ] REVISION of application number: ____________________.
(This application replaces a prior version of a new, competing continuation or supplemental application.)

[ ] Change of Principal Investigator/Program Director.
Name of former Principal Investigator/Program Director: ____________________.

ASSURANCES IN CONNECTION WITH:

Civil Rights	Handicapped Individuals	Sex Discrimination	Human Subjects General Assurance *(If applicable)*	Laboratory Animals *(If applicable)*
[ ] Filed	[ ] Filed	[ ] Filed	[ ] Filed	[ ] Filed
[ ] Not filed	[ ] Not filed	[ ] Not filed	[ ] Not filed	[ ] Not filed

INDIRECT COSTS:

Indicate the applicant organization's most recent indirect cost rate established with the appropriate DHHS Regional Office. If the applicant organization is in the process of initially developing or renegotiating a rate, or has established a rate with another Federal agency, it should, immediately upon notification that an award will be made, develop a tentative indirect cost rate proposal based on its most recently completed fiscal year in accordance with the principles set forth in the pertinent *DHHS Guide for Establishing Indirect Cost Rates*, and submit it to the appropriate DHHS Regional Office. Indirect costs will not be paid on foreign grants, construction grants, and grants to individuals, and usually not on grants in support of conferences.

[ ] DHHS Agreement Dated: ____________________.
__________ % Salary and Wages or __________ % Total Direct Costs.

Is this an off-site or other special rate, or is more than one rate involved? [ ] YES [ ] NO
Explanation: ____________________

[ ] DHHS Agreement being negotiated with ____________________ Regional Office.

[ ] No DHHS Agreement, but rate established with ____________________ Date __________

[ ] No Indirect Costs Requested.

PHS-398
Rev. 5/80

PRINCIPAL INVESTIGATOR/PROGRAM DIRECTOR: Lewis, Bryan A.

DETACH AND CLIP TO THE SIGNED FACE PAGE OF THE APPLICATION

PERSONAL DATA ON

PRINCIPAL INVESTIGATOR/PROGRAM DIRECTOR

The Public Health Service has a continuing commitment to monitoring the operation of its review and award processes to detect–and deal appropriately with–any instances of real or apparent inequities with respect to age, sex, race, or ethnicity of the proposed principal investigator/program director.

To provide the PHS with the information it needs for this important task, the principal investigator/program director is requested to complete the form below and attach a single copy to the signed face page of the application.

Upon receipt and assignment of the application by the PHS, this form will be detached from the application. It will NOT be duplicated and will NOT be a part of the review process. Data will be confidential, and will be maintained in Privacy Act record system 09-25-0036, "Grants: IMPAC (Grant Contract Information)." All analyses conducted on the data will report aggregate statistical findings only and will not identify individuals.

If you decline to provide this information, it will in no way affect consideration of your application.

Your cooperation will be appreciated.

Date of Birth: 05/23/51 (Month/Day/Year)   Sex: [ ] Female [X] Male

Race and/or Ethnic Origin:

Check one:

- [ ] American Indian or Alaskan Native
- [ ] Asian or Pacific Islander
- [X] Black, not of Hispanic origin
- [ ] Hispanic
- [ ] White, not of Hispanic origin

NOTE: The category that most closely reflects the individual's recognition in the community should be used for purposes of reporting mixed racial and/or ethnic origins. Definitions are on the back of form.

B. ANOTHER EXAMPLE OF AN APPROVED AND FUNDED PROJECT: REDUCING MEDICATION ADMINISTRATION ERRORS IN FAMILY CARE HOMES THROUGH A PROGRAM OF ON-SITE EDUCATION AND TECHNICAL ASSISTANCE

THE UNIVERSITY OF NORTH CAROLINA

ABSTRACT OF APPLICATION FOR GRANT, CONTRACT, OR COOPERATIVE AGREEMENT

(See instructions on reverse.)

Title: Reducing Medication Administration Errors in Family Care Homes Through a Program of On-Site Education and Technical Assistance

Number: 00

Principal Investigator/Project Director: Abraham G. Hartzema

The scant existing literature available indicates that drug management problems are prevalent in board and care homes (Hartzema, 1985) and contribute to adverse patient outcomes (Budden, 1985), but causes contributing to those problems are not well researched. The major thrust in this project will be in promoting effective ways to overcome the weaknesses in drug therapy management in Family Care Homes.

The proposed program of on-site educational and technical assistance will focus on record keeping practices and medication administration procedures in a sample of 96 North Carolina family care homes. In an experimental design 72 Family Care Homes will receive the intervention while 24 facilities will serve as controls. The intervention entails the implementation of a medical chart system appropriate for the resources available in the facilities, review of existing medication administration procedures and guidance in PRN (as needed) medication administration decisions.

Persistence of the intervention will be evaluated by measuring medication administration errors, using techniques proposed by the American Society of Clinical Pharmacists (Medication Error Survey, 1984), at selected time intervals, and comparing the observed medication administration error rates with those observed in the control group. Observed medication administration errors will be related to sources of error in medication administration record keeping and medication administration procedures.

Three training programs will be organized to instruct county social workers in charge of oversight of these facilities in techniques and methods to identify sources of medication administration errors, remediate these errors and assure persistence in good medication administration record keeping practices and procedures in Family Care Homes.

Reducing Medication Administration Errors in Family Care Homes Through A Program of On-Site Educational and Technical Assistance

Summary Description

The scant existing literature available indicates that drug management problems are prevalent in board and care homes (Hartzema, 1985) and contribute to adverse patient outcomes (Budden, 1985), but causes contributing to those problems are not well researched. The major thrust in this project will be in promoting effective ways to overcome the weaknesses in drug therapy management in Family Care Homes.

The proposed program of on-site educational and technical assistance will focus on record keeping practices and medication administration procedures in a sample of 96 North Carolina family care homes. In an experimental design 72 Family Care Homes will receive the intervention while 24 facilities will serve as controls. The intervention entails the implementation of a medical chart system appropriate for the resources available in the facilities, review of existing medication administration procedures and guidance in PRN (as needed) medication administration decisions.

Persistence of the intervention will be evaluated by measuring medication administration errors, using techniques proposed by the American Society of Clinical Pharmacists (Medication Error Survey, 1984), at selected time intervals, and comparing the observed medication administration error rates with those observed in the control group. Observed medication administration errors will be related to sources of error in medication administration record keeping and medication administration procedures.

Three training programs will be organized to instruct county social workers in charge of oversight of these facilities in techniques and methods to identify sources of medication administration errors, remediate these errors and assure persistence in good medication administration record keeping practices and procedures in Family Care Homes.

[redacted] Foundation
PROJECT PERIOD: OCTOBER 1, 1989 — FEBRUARY 28, 1991 (17 MONTHS)
A. HARTZEMA, PRINCIPAL INVESTIGATOR

BUDGET CATEGORY

PERSONNEL	ROLE IN PROJECT	PERCENT EFFORT	SALARY	FRINGES	TOTAL	
A. Hartzema	Principal Investigator	10%				
P. Sloane	Co-Investigator	10%				
G. Koch	Co-Investigator	10%				
C. Hogue	Co-Investigator	10%				
H. Lewis	Accounting Manager	10%				
TBA	Pharmacist	100%				
TBA	Research Assistant	50%				
TBA	Secretary	50%				
TOTAL PERSONNEL			132,001	27,895	159,896	159,896
TRAVEL						
In-state mileage to 96 homes × 200 mile radius × 2 sessions					9,600	
Three workshops for county social workers (3 × 200 mile radius × $.25/mile					150	9,750
EQUIPMENT						
HPLaser Printer Series II					1,290	1,290
SUPPLIES						
General Office Supplies (stationary, computer paper, etc.)					800	
Pharmacy/Nursing Home Forms (Med. Adm. Record, Tx Record, Nurses' Notes, etc.)					206	1006
OTHER						
Data Entry, Management and Analysis: $1,000 mainframe plus 350 hrs × $20					8,000	8,000
TOTAL DIRECT COST						179,942

TOTAL DIRECT COST TO ADMINISTRATION ON AGING	92,600	92,600
INDIRECT COST @ 8%	7,400	
TOTAL COST ADMINISTRATION ON AGING	100,000	
TOTAL DIRECT COST TO [redacted] FOUNDATION	87,342	
NO INDIRECT COST		
TOTAL COST [redacted] FOUNDATION		87,342
		179,942

REDUCING MEDICATION ADMINISTRATION ERRORS IN FAMILY CARE HOMES THROUGH A PROGRAM OF ON-SITE EDUCATIONAL AND TECHNICAL ASSISTANCE

a. OBJECTIVES AND NEED FOR ASSISTANCE

The proposed study will examine a structured program of educational and technical assistance to promote improved medication administration record keeping practices, medication administration procedures and PRN (as needed) medication administration decisions in family care homes. The long term objectives of the project are to reduce medication administration errors and consequently prevent adverse or suboptimal drug therapy outcomes in the elderly residing in family care homes.

Aging and Drug Therapy Management Issues in Board and Care Homes. The impact of recent and anticipated demographic changes on the health care requirements of the impaired elderly is one which is only now becoming apparent, but which will assume increasing economic and social importance well into the next century. Over the coming decades, our elderly population faces a diminishing availability of informal family supports, an increasingly complex and costly health care delivery system, along with the physical vulnerability that comes with the aging process.

The elderly are predisposed to experience therapeutic problems with drugs not only because of the biological effects of aging, but also because of their pattern of drug consumption and the kinds of limitations they experience in their health care environments. The aging process results in physiological changes which, in time, affect the absorption, distribution, and metabolism of medications (Hollister, 1981), resulting in a ''narrow therapeutic window'' where ''the range between dosage levels for benefits and side effects narrows for older patients'' (Kane, 1985). As a result, the response of elderly patients to drug therapy is often unpredictable and requires careful drug selection, dosing, administration and monitoring.

Secondly, the elderly show a high prevalence of chronic illness and physical impairment (Vener, 1979), entailing substantial drug utilization. Although they comprise only 11 percent of the U.S. population, the elderly account for 25 percent of drug expenditures (Gibson, 1977). Recent studies estimate that among the elderly aged 75 or older, approximately 87 percent of the ambulatory and 95 percent of the institutionalized take at least one prescribed medication (Law, 1976). Further, polypharmacy is high among the elderly (Kalchthaler, 1977). For residents in North Carolina board and care homes taking at least one prescribed medication, 5.8 drugs are concurrently prescribed (Hartzema and Konrad, 1985). Because concurrent consumption of several medications increases the probability of adverse drug reaction (ADR) experiences, it is not surprising that studies have documented the high frequency of adverse effects of prescription drugs in the elderly (Lamy, 1981). Budden (1985) found a 26 percent incidence of ADRs in residents of board and care homes admitted to an acute geriatric ward in a general hospital.

Certain drug side effects are particularly significant. Sedatives producing orthostatic hypotension may cause falls which can result in fractures (Ray, 1987). Gastrointestinal side effects, such as appetite depression, nausea, and stomach irritation, can be particularly important in the elderly whose nutritional status may already be compromised. Glaucoma (Godwin-Austen, 1971), cholestatic jaundice (Holloway, 1974), hyperkalemia (Meier, 1983), cognitive impairment (Larson, 1987) and tardive dyskinesia (Goetz, 1982; Jenke, 1983) may also be produced or exacerbated by medication. This situation is complicated by the fact that practicing physicians serving elderly patients may have limited knowledge about the special drug needs of this age group (Ferry, 1985; Dyer, 1984) or as is the case in board and care homes have little involvement in patient care (Avorn, 1989).

Finally, the elderly experience a high incidence of medication administration errors. Estimates of the incidence of such errors in the elderly population range from 25 percent to 50 percent (Parkin, 1976; Schwartz, 1962). In nursing home facilities, estimates have been as high as 59 percent (Morse, 1982); in a pilot study conducted in five North Carolina board and care homes, error rates of 104% were found (Hartzema, 1986). While trained personnel such as licensed nurses, aides, or medication technicians are responsible for medication administration in nursing homes, errors may be compounded in board and care homes where persons with limited training are in charge of medication administration. Since many such personnel are elderly themselves, with limited formal education and medical background (Sherwood, 1983; Bradshaw, 1976), who experience difficulties in reading and understanding the prescribed drug regimens (Nelson, 1988), therefore, errors in medication administration are likely to occur in the absence of special efforts to prevent them.

Board and care homes or domiciliary care facilities (DCFs) are essentially quasi-health care institutions, staffed by lay persons. Scanlon (1983) has observed that these institutions are specifically excluded from providing ''nursing care'', but are mandated ''to provide...protective oversight to their residents'', including ''central storing and distribution of medication, assistance in taking medication, [and] assistance with acquiring medical and dental care.'' (Emphasis added.) Lay caregivers are thus placed in a double-bind by society, explicitly prohibited from being identified as health care providers, yet expected to carry out and be held accountable for the performance of fairly complex health-related tasks and responsibilities. These demands certainly require that caregivers have substantial familiarity with, and significant participation in, medical decision making; but even while avoiding a decisive role in medical or pharmaceutical actions, caregiver initiative is required to coordinate physician, pharmacist, family, and resident information, communication and interaction.

Given such an ambiguous mandate, and because the agencies overseeing board and care homes in most states tend to be less comfortable with medical than with social welfare issues, it is not surprising that drug management was identified in a nationwide study of domiciliary care as the first ranking quality of care issue (Stone, 1985).

The Drug Therapy Management Process in Board and Care Homes. A key feature of the drug therapy management process in board and care homes is a circular flow of information and control involving several roles: the prescribing physician, the dispensing pharmacist, the caregiver(s) in the board and care home, and the drug regimen reviewer (DRR) — generally a consulting pharmacist. The communication between physician and dispensing pharmacist takes place largely through the prescription. The medication label and physician orders in the family care home should jointly contain sufficient information for the caregiver to administer the right medication at the right time to the right patient. If correct information is available at the board and care home and contained in the medical chart of the patients, proper drug administration decisions can be made. Review of the drug therapy management process in family care homes suggests that substantial improvement can take place in medication administration record keeping practices, medication administration procedures and PRN decisions.

Preliminary Studies. Several studies of the pattern of drug therapy management in board and care homes have been conducted by the investigators, their University of North Carolina colleagues, the Long Term Care Association, and various state agencies. Prescribing patterns were assessed in the investigators' examination of one month's pharmacy records in a sample of 10 percent of all SNF, ICF, and board and care homes in North Carolina. While SNF residents had more drugs prescribed (8.2) than residents of ICF (7.6) or residents of board and care homes (5.8), the latter are still relatively high drug users and prone to the risks of such use. The nature of the drugs used by board and care homes residents shows a similarity with the drugs used in nursing homes. Five categories — sedatives, hypnotics, analgesics, cathartics, and tranquilizers — accounted for the majority of all prescribed drugs. Many drugs are prescribed on a PRN (take as needed) basis (32 percent of legend drugs and 69 percent of OTC drugs) so that staff of board and care homes with minimal medical training must make therapeutic decisions. The quantity of drugs prescribed, their nature, and the large percentage of drugs (such as Haldol) on a PRN basis confirm the need for adequate medication record keeping practices and drug administration procedures in family care homes (Hartzema and Konrad, 1985; see Appendix A).

Drug administration procedures were addressed in a survey of 68 large board and care homes in North Carolina in 1971. Only larger ones maintained medication records, 33 percent had no dedicated drug storage place, and in 69 percent employees had only on-the-job training (Eckel, 1971). A more recent, state-sponsored medication control survey of 30 self-selected North Carolina facilities identified problems including mislabeling, mixing of external and oral medications, no accountability of controlled substances, no vital signs to support drug therapy, and incomplete medication cards as the major problems (see Appendix B).

In a master's thesis research project conducted by Lee (1987), 105 board and care managers were interviewed to obtain information on the organization of pharmaceutical services, record keeping practices, and telephone orders. Thirty-eight percent of the respondents used a computerized medication administration record (MAR) provided by the pharmacist, 26% used a form supplied by the Department of Social Services, and 20% used their "own" form, while the remaining managers used the FL2 (a Medicaid utilization review form) as basis for medication administration. Telephone prescribing occurred frequently; 29.5% of all drug orders were phoned in to the pharmacy by board and care home personnel (caregivers), but frequently no procedures were established to enter these orders in the Medical Charts. Written physician orders are an important source of information for caregivers, 86.7% of the residents' medical charts accompanied the resident to the physician, however in many cases these charts were not complete.

The feasibility of an educational intervention dealing with drug distribution and the nature of the target population were addressed in a telephone survey of 20 randomly selected rest homes conducted by the investigators. Although all 20 rest homes recorded PRN drug administrations, only half documented the indications on which the drugs were given, and only two of the 20 facilities had written in-house drug administration guidelines or policies. Further, more than 80 percent of the administrators were willing to participate in an on-site educational intervention to improve medication record keeping practices and review medication administration procedures (see Appendix C).

Pilot studies (Hartzema, 1985) indicated that at least 30% of the prescription medication in board and care facilities are written on an as needed basis (PRN), thus offering operators of family care homes little guidance in the use of these medications. White (1988) found that board and care home operators often have little guidelines for the administration of PRN medication and recommends the development of methods to communicate PRN indications to appropriate board and care personnel.

Dr. Hartzema evaluated the effect of a structured educational intervention on medication administration procedures and outcomes in board and care homes during 1986. In a quasi-experimental design, on-site observations of medication administration were conducted in five rest homes, pre- and post-intervention. All board and care homes selected to participate agreed to enroll in the study. Based on the sources of medication administration errors observed in the pretest, a structured 8-hour problem-oriented educational intervention was developed and implemented. Transcription errors (38.5%), unauthorized medications (18.7%), prescribed medications not available

TABLE 1
Percent Error Rate in Study Board and Care Homes Before and After Educational Intervention

Facility*	Pre Test Percent Error Rate**		Post Test Percent Error Rate**	
A	175.0	(49/28)	108.0	(54/50)
B	74.4	(29/39)	85.4	(35/41)
C	96.7	(29/30)	38.5	(10/26)
D	200.0	(50/25)	168.0	(42/25)
E	280.0	(14/5)	233.3	(14/6)
Average	134.6	(171/127)***	104.7	(155/148)***

* Facility A 85 licensed beds, B 57 licensed beds, C 20 licensed beds, D 5 licensed beds and E 5 licensed beds.

** Total drug (administration errors/total opportunity for error) × 100. Error rates over 100% can be observed if multiple errors are made for a dose administration, such as wrong time and wrong route for same dose administered.

***t = 2.8 (4 d.f.), for a one-tailed test $p < .025$.

(19.1%) and omissions (9.9%) were the most commonly observed errors. Post-intervention observations found a decrease in medication administration errors; medications errors decreased significantly from 134.6% to 104.7% ($p < 0.25$) for all facilities (see Table 1).

Two types of board and care homes were enrolled in the study. The larger homes (A, B, and C; see Tables 1 and 2), licensed for seven or more adults, are referred to as Homes for the Aged, while Family Care Homes (D and E; see Table 1 and 2) are licensed for two to six adults. The Family Care Homes showed a higher overall medication administration error rate than the larger Homes for the Aged. Transcription errors, errors introduced by the monthly transcribing of past months information to new medication administration records, introduced a large number of such errors. Family Care Homes showed a higher overall medication administration error rate as well as a high transcribing error rate, while the classroom education intervention resulted in only little effect on the reduction of transcribing errors. Moreover, transcribing errors may contribute to other observed errors such as wrong dosing, wrong time or omission of the drug. Erroneous medication administration records caused by transcription mistakes are a major cause of deviation of drug therapy from physician orders.

This study demonstrated the success in recruiting the board and care homes, the feasibility of the proposed medication administration survey methodology. However, the educational intervention appeared only to be marginally effective and the authors suggested the evaluation of a program of on site technical assistance and education to reduce the number of errors introduced by the monthly transcription of medication administration records in Family Care Homes. The socio-demographic composition of the caregiver population, which are often elderly females themselves, and include a larger percentage of minorities than the U.S. population, and the difficulty faced in the smaller Family Care Homes to find replacement when participating in outside educational activities underscore the validity of a program of on-site educational and technical assistance. See Appendix D for selected data of this study.

A state-wide survey of board and care homes conducted by the continuing education director of the North Carolina Long Term Care Association found that administrators ranked drug therapy issues as a top priority, especially understanding drug terminology, procedures, charting and accountability, disposal of discontinued drugs, and filling and dispensing of drugs (Floren, 1985). Recently instituted mandatory continuing education (CE) requirements for administrators have heightened interest in medication topics (see Appendix E).

b. RESULTS AND BENEFITS EXPECTED

The results and benefits of the proposed program are manyfold, and some can be qualified as direct outcomes of the program while other outcomes of the program are indirect.

The direct outcomes of the program are a reduction in medication administration errors achieved by a proper medical chart system, correct medication administration procedures and sound PRN decisions. A reduction in medication administration errors will contribute to drug therapy in accordance with physicians' orders and consequently may contribute indirectly to an improved cognitive, physical and psycho-social functioning of the elderly residing in Family Care Homes.

Other direct outcomes are an increase in our knowledge concerning *sources* of medication error in these facilities. Knowledge concerning sources of error allows us to design strategies to prevent medication administration

TABLE 2
Medication Administration Errors by Type of Error and Facility

	Pre Test						Post Test					
	A	**B**	**C**	**D**	**E**	**Total**	**A**	**B**	**C**	**D**	**E**	**Total**
Omissions	3	0	4	8	2	17	21	0	5	3	2	31
Unauthorized Drugs	13	2	4	9	4	32	8	7	1	7	4	27
Wrong Dose	4	6	0	0	3	13	1	1	0	0	2	4
Wrong Route	0	0	0	0	0	0	0	0	0	0	0	0
Wrong Dose Form	0	0	1	0	0	1	0	0	1	0	0	1
Other	29	21	20	33	5	108	24	27	3	32	6	92
Time	3	1	1	0	2	7	1	0	0	0	2	3
Trans	19	9	12	25	1	66	20	23	3	27	1	74
No Drug Avail	7	11	4	7	2	31	3	4	0	4	2	13
Not Observed	0	0	2	0	0	2	0	0	0	0	1	1
Wrong Indicat.	0	0	1	1	0	2	0	0	0	1	0	1
Total Errors	49	29	29	50	14	171	54	35	10	42	14	155
MD Error	0	0	1	4	1	6	2	0	1	3	1	7
Total Opport. for Error	28	39	30	25	5	127	50	41	26	25	6	148

Facility A 85 licensed beds,B 57 licensed beds, C 20 licensed beds, D 5 licensed beds and E 5 licensed beds.

errors. Moreover, materials, techniques and training programs will be developed during the program focussed at the major sources of medication administration errors. These materials, techniques, procedures and training programs will be shared with North Carolina's Department of Health and Human Services, Division on Aging and Division of Social work, the North Carolina Longterm Care Association and the National Association of Residential Care Facilities. These agencies and associations have committed themselves to further disseminate the tangible products of the program. See Appendix F for letters of commitment.

County social workers bear the responsibilities to assure quality of care in board and care homes. Social workers due to the focus of their training have in general more affinity with social issues than with health care issues. To introduce social workers to medication administration quality assurance techniques, three workshops will be organized within the framework of the proposed program. It is anticipated that additional seminars will be organized and the investigators certainly will stimulate and contribute to these seminars after completion of the program.

c. APPROACH

The Setting. Board and care represents an increasingly significant resource in the long term care service continuum. An estimated 400,000 to 800,000 beds are available (Mellody, 1979; Stone, 1985; Mor, 1986; Jameson, 1988), which mainly service an elderly chronically impaired population (Stone, 1985; Nelson, 1988). The low visibility of the board and care industry no doubt stems in part from the status attributes of the population it services, which can be characterized as predominantly an elderly, chronically mentally ill, female, low income, and minority population. However, the national scope and size of this long term care option is substantial, meriting considerably more research attention than it has received to date (Mor, 1986).

North Carolina's supply of board and care homes include homes for the developmentally disabled adults, homes for the aged, and family care homes, with a total of 20,240 licensed beds. This compares to 21,213 nursing homes beds in the state, some 10,766 in ICFs and 10,447 in SNFs. Located in the State of North Carolina are 631 licensed Family Care Homes which house approximately 3,381 residents (DHHS-NC, 1989). The Domiciliary Home Survey Annual Population Report 1982—1983 indicates that 76.4 percent of the board and care population falls in the age group 60 and older (see Appendix G). Nelson (1988) found in a survey of the care needs of the residents of board and care homes in a North Carolina county, that the mean age of the residents was 71.5, with 70% over the age of 65 and 20% over the age of 85. Psychotropic drugs were used by 58% of the board and care residents.

Revision of current N.C. standards for board and care homes became effective in 1985. (See Appendix H for September 1987 standards.) These standards describe regulations concerning the storage requirements of the medications, administration of medications, and maintenance of records. Most importantly, the standards made the managers responsible for medication administration, and for interacting with physicians on their residents' behalf. Since North Carolina is a state which leads in its use of board and care within the spectrum of long-term care and its range of regulatory and enforcement policies is consistent with the emerging experience of other states, it is expected that this proposed program of on-site educational and technical assistance for Family Care Homes program will come to have increasing national relevance.

Research Plan. In the first phase, family care homes will be selected, participation will be secured and arrangements will be made for on-site observations.

In the second phase, informed consent of all administrators will be obtained and on-site observations of medication administration will be performed at all family care homes employing a standard Medication Administration Error Survey Technique (ASCP, 1984). Pre-intervention data will be analyzed, diagnosis will be made and an action plan for technical assistance will be developed. Record keeping systems and proper pharmacy administration procedures will be introduced in the experimental group of family care homes, while as alternate, a pharmacology program will be provided in the control homes. Again, direct after the implementation of the intervention observation of medication administration errors will be performed in experimental as well as control homes.

In the third phase, post-intervention measurements will again be performed at all facilities. Medication administration will be surveyed and the quality of medical charting assessed to evaluate the persistence of the intervention. In the control facilities, at this time, proper record keeping systems will be introduced, medications administration practices reviewed and guidance provided in proper PRN decision making techniques.

In the fourth phase, the evaluation results will be analyzed and reported, intervention techniques will be improved, and materials, techniques, and procedures will be shared through three seminars with county social service workers.

Phase I

Recruitment of family care homes: The overall evaluation design requires the selection of 72 experimental and 24 control facilities in order to assess the impact of the intervention components on the outcomes. Geographic clustering of the family care homes is desirable (1) to minimize administrative and travel costs; (2) to increase replicability by stimulating regionalized continuing education efforts; and (3) to reduce the possibility of contamination between Experimental and Control groups that might be selected with the same physicians, pharmacists, or social workers.

North Carolina's 100 counties will be categorized according to size, rural counties with less than 100,000 inhabitants and urban counties with over 100,000 inhabitants, and according to east and west regions. After selection, counties will be randomly assigned to either the experimental or control group. The experimental group will receive the technical assistance and pre- and post-measurements, whereas the control group will receive the pre- and post-measurements and an alternate intervention. Sampling considerations are discussed in more detail in the section "Geographic Location".

Family care home recruitment will follow the approach as outlined by Palumbo (Palumbo, 1987). Briefly, this approach entails an initial phone contact, followed by an on-site visit by investigators to explain the purpose of the study and provide evidence of support from long term care trade associations, professional associations, and state officials. Informed consent will be obtained prior to assessing the medication administration system in the facilities. This approach achieved a successful recruitment rate of 87% of the facilities sampled; among board and care homes, a 100% recruitment was achieved. A 100% recruitment rate was also found in our pilot study (Hartzema, 1986). Participation of family care homes will be further assured through the agency of the state and local social service departments and the North Carolina Long Term Care Association. (See letter of support in Appendix F.) Principal investigator (Dr. Hartzema) will contact the family care homes and assure participation.

Phase II

The next phase of the project involves a site visit by project staff (a pharmacist) to each of the 96 family care homes. At that time several tasks will be accomplished.

(1) Informed consent of administrators, caregivers, and residents will be obtained.
(2) Medication profiles for residents of the family care homes will be prepared.
(3) Family care home record keeping practices will be reviewed.
(4) Medication administration will be unobtrusively observed during the morning round to record medication administration error rates and identify sources of error.

Problem Identification: Medication Error Observation. The pharmaceutical services in both experimental and control board and care homes will be examined, and on-site observations will be made of medication administration procedures. Collection of medication administration error data will rely on a modified Medication Administration Error Survey (MAES) which employs a disguised observation technique. The MAES measures "the administration of dose of a drug or biological that deviates from the physician's order contained in the patient's medical record" (HCFA, 1985), and is a refinement of Barker's (1962, 1969, 1982, 1984) technique now recommended by the American Society of Consultant Pharmacists (ASCP, 1984). As an outcome measure of the drug distribution process, the MAES detects errors at the point of drug administration and classifies them by sources of error (see Appendix I).

Two pharmacist observers will be trained in the MAES techniques. One project observer will be used in site visits to the family care homes. The observer will document the medication administration practice of the rest home caregiver who is assigned the responsibility of drug administration during morning administration for the residents. The same observer will be employed to observe the same caregiver pre- and post-test intervention. The feasibility of this approach has been demonstrated in a pilot study (Hartzema, 1986).

Phase III

Educational Intervention. Systematically collected information provided by the medication error observer will be used for the needs assessment. These needs and priorities will be reduced to technical assistance objectives; accordingly these objectives will be implemented.

The specific objectives for the interventions include:

1. development of systematic procedures of administration medications;
2. implementation of appropriately detailed yet feasible methods for recording the actual administration of medications and organizing patient information in medical charts;
3. introduction of structured methods for transcribing medication administration records and verification of transcribed records; and
4. adoption of a check list to abstract PRN medication administration information necessary for appropriate decisions such as indication for use, dosing schedules, length of administration and stop indications from physicians orders (PRN decision protocols).

Attainment of objectives for educational and technical assistance:

1. Implementation of appropriate detailed yet flexible methods for recording the actual administration of medications.

Owners, operators or administrators of the family care homes will be assisted in developing medical charts for each patient. These medical charts will be composed of

a. physician notes.
b. progress notes.
c. medication administration records (MAR).
d. FL2 forms.
e. drug regimen review notes.
f. PRN decision protocols.

If charts are not available or if the only chart system available is the "yellow pad", a chart system and a supply of record forms will be provided with a sufficient supply for a three month period. In addition information will be provided about the availability and supplies of these record forms.

2. Development of systematic procedures of administration medications.

Based on an analysis of errors observed during the morning round, strategies to prevent these errors will be discussed with the home operator in a non-threatening atmosphere. During the preparation of the evening medication round the following medication administration round, each step will be reviewed and advice and assistance will be provided in approaches and techniques which will limit the occurrence of such errors.

3. Introduction of structure methods for transcribing medication administration records and verification of transcribed records.

The following procedures and steps in MAR transcribing will be introduced:

a. MAR will be transcribed to new form;
b. transcriber verifies new MAR with old MAR;
c. second person verifies new MAR with old MAR;
d. transcriber verifies data elements on prescription; container label with MAR entries;

e. discrepancies between Rx container labels will be examined by:
   i) review of physician orders;
   ii) phone call to pharmacy for clarification and/or
   iii) phone call to physician for clarification.

4. Adoption of a checklist to abstract PRN medication information necessary for appropriate administration decisions, such as indication for use, dosing schedules, length of administration and stop indications from physician orders and prescription container label.

Identify the two sources of information important to complete PRN checklist; these two sources are physician orders and prescription container labels. Information to be abstracted for the PRN checklist includes:

PRN Checklist	Source of Data
a. Drug name	Rx label + physician order
b. Indications for use	Physicians orders + Rx label
c. Symptoms to look for/when not to use	Physicians orders
d. Dosing required	Rx label
e. Length of therapy (maximum)	Physicians orders
f. Evaluation of symptoms and how to evaluate drug effects	Physicians orders
g. Side effects to expect	Physicians orders
h. What to do on observation of side effects	Physicians orders
i. What action to take when symptoms continue	Patient to physician or emergency room

The technical assistance program will include such activities as the organization of medical charts for the residents, advise about the selection of appropriate forms, and consultation on proper medication administration procedures. These activities will take place concurrently, but after the pre-intervention observations are made, and before the post-intervention observations are performed. Post-intervention observations will be made on the same day following the interventions, to evaluate in the experimental facilities the direct impact of the intervention and in the control facilities to assess the variability in the observations.

Phase IV

In the evaluation stage, the medication administration procedures will be reviewed employing similar observation and documentation techniques as employed in the pre-intervention phase. Persistence of the changes made during the on site educational and technical assistance in medication administration practices and procedures will be evaluated by observing 24 experimental facilities after a three month period interval, 24 experimental facilities after a six month period interval and 24 experimental facilities after a nine month period interval. Corrections and adjustments deemed necessary will be made and barriers to persistence of the implemented charting procedures and administration policies will be identified as well as factors that contribute to the persistence of the effects over time. See Figures 1 and 2 for a presentation of the study design.

Evaluation of Drug Administration Intervention. Evaluation of the efficacy of the intervention will be conducted by comparing before-after and experimental-control groups on the incidence of medication administration errors (i.e. the total number of errors over the number of total opportunities for error (TOEs) multiplied by 100 percent) (ASCP, 1984). In addition, percentage error rates will be calculated for separate categories of errors, such as omissions, unauthorized drugs, wrong doses, wrong route, wrong dose form, and other errors. These categorical error rates will assist in identifying the sources of error that are adequately reduced by the educational intervention and technical assistance and the sources of error which require additional attention in further development of programs for educational and technical assistance for drug administration procedures in family care homes.

Evaluation of Drug Administration in Family Care Homes. Evaluation outcomes and indicators will be of several types:

Main Outcome:

1. Number of medication administration errors observed per family care home.

Indicator for Main Outcome:

1. Percent error rate calculated as the total number of observed errors over the number of doses administered and omitted, multiplied by 100.

Secondary Outcomes:

1. Medication administration errors by category observed per family care home.
   a. transcription errors and errors introduced through transcription errors:
      i) omissions observed
      ii) unauthorized drugs observed

TIME LINE

1989 Oct	Nov	Dec	1990 Jan	Feb	Mar	Apr	May	Jun	Jul	Aug	Sep	Oct	Nov	Dec	1991 Jan	Feb

VISIT SCHEDULE OF FACILITIES

	Experimental Group Before						Experimental Group After						
Random Selection of Family Care Homes	A_{12}	B_{12}	C_{12}	E_{12}	F_{12}	D_{12}	E_{12}	F_{12}	C_{12}	A_{12}	B_{12}	D_{12}	Report Writing
													Preparation of Articles

	Control Group Before			Control Group After		
Preparation of Intervention	X_8	Y_8	Z_8	X_8	Y_8	Z_8

Development Instruments

Recruitment of Personnel

Homes		Evaluation Period
A B	-	9 months
C D	-	6 months
E F	-	3 months
X Y Z	-	9 months

One facility/day/pharmacist
two 50% pharmacists
5 facilities/week
20 per month

FIGURE 1.

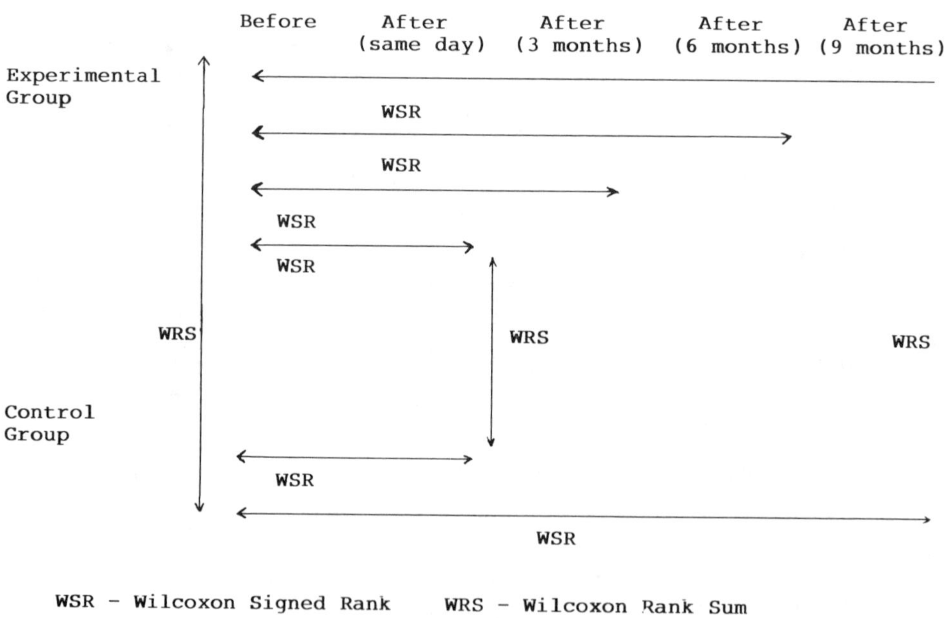

FIGURE 2. Family care home intervention.

iii) wrong dose observed
iv) wrong route observed
v) wrong dose form observed
vi) other errors observed

b. errors not caused by transcription errors:

2. Completed PRN administration decisions protocols for the residents.
   i) omissions observed
   ii) unauthorized drugs observed
   iii) wrong dose observed
   iv) wrong route observed
   v) wrong dose form observed
   vi) other errors observed
3. Completeness of medical charts for each family care home.

Indicators for Secondary Outcomes:

1. Categorical percent error rate calculated as the number of errors in the error category over the number of doses administered and omitted multiplied by 100.
   i) omissions percent error rate
   ii) unauthorized drug percent error rate
   iii) wrong dose percent error rate
   iv) wrong route percent error rate
   v) wrong dose form percent error rate
   vi) other percent error rate
2. Rate of PRN administrations decision protocols with complete information over the number of medications prescribed on a PRN basis.
3. Rate of correct completion of forms.

Sample size considerations. Sample size considerations are based on a pilot study which was conducted in North Carolina rest homes by the principal investigator using related methods. In this study employing a quasi-experimental design with pre-intervention medication administration observation in five facilities, it was found that an educational intervention could reduce the number of medication errors by 30%. The found reduction in error rate was significant at the $p < .025$ level. (See Appendix D.) The goal of this study will be to reduce the medication administration error rate in family care homes by at least 30%.

Table 1 shows that the differences between pre- and post-tests range from −67 to +11 for the pilot study. For convenience, let us assume that the range is from −80 to 20; then the range is 100 units. The range divided by 4 provides us with an approximate standard deviation (δ) of 25 units. In the proposed study a total of 96 facilities will be enrolled, 72 experimental facilities and 24 control facilities. Most analyses will be performed in terms of comparisons between a group of 24 control and 24 experimental facilities or 24 experimental facilities and 24 experimental facilities observed at different time intervals. For an alpha equaling .05, with a two sided test and a power of .80, the following formula can be applied:

$$\Delta^2 = \frac{(Z\alpha + Z\beta)^2\, 2(\sigma^2)}{24} = \frac{(1.96 + 0.84)^2\, 2(625)}{24} = 408.3$$

$$\Delta = \sqrt{408.3} = 20.2\%$$

= difference we will be able to detect for sample size of this order. Thus, in the proposed study, we will be able to detect a change of 20% or more in medication administration error rates.

Statistical Analysis. In this section, some of the statistical analyses which will be performed on the main outcome are outlined. Many analyses not described here will be generated as the findings evolve. Also, in all analyses, the family care home is the unit of analysis.

The principal objective of statistical analysis for the main outcomes for the proposed study are: (I) comparisons of the control and experimental group of family care homes for the incidence of medication error rates pre- and post-intervention at the day of the intervention; (II) comparison of the incidence of medication errors rates before and after over a three month time period, a six month time period and a nine month time period to evaluate the persistence of the intervention; (III) comparison of the control group and the experimental group (nine month interval) post-intervention; (IV) comparison of the intervention group and control group for antecedent variables such as region, population size, etc. See Figure 1 for the timeline and the study design.

The analyses of the interventions entails the use of the Wilcoxon Signed Rank test for within group comparisons of pre-intervention scores and post-intervention scores to evaluate the effects of the technical assistance. The Wilcoxon Rank Sum test will be used for between groups comparison for scores before and after the intervention and for the changes between pre-test and post-test. If prescores or other background variables need to be taken into account, rank analysis of covariance methods will be used for between group comparison. Further analysis may employ multivariate statistical models to describe the relationship of outcome measures to a variety of explanatory variables that apply at the family care home level. See Figure 2.

Patterns of association among alternative outcome criteria at pre-intervention and post-intervention will be evaluated for the respective groups with Spearman rank correlation coefficients. Such methods will also be applied to changes between pre-intervention and post-intervention.

d. GEOGRAPHIC LOCATION

Sample Frame Consideration. North Carolina's supply of rest homes includes 631 Family Care Homes licensed for 2—6 residents. Family Care Homes are located in counties of different size. Two strata are of importance to this study because they reflect the demographic structure of counties nationwide as shown by census data. These are the rural counties with less than 100,000 inhabitants, and the urban counties with 100,000 inhabitants and over.

North Carolina is composed of 100 counties; Family Care Homes are located in 82 counties of which 67 counties have a population less than 100,000, and 15 counties show a population census of 100,000+ (North Carolina Office of State Budget, 1988). In addition the counties are categorized by the two regions: east and west. See Table 3 for a listing of counties, population size, geographic area, and number of Family Care Homes.

A two stage sample design is developed for randomly selecting the Family Care Homes. In the first phase of the design, counties are randomly selected based on two criteria, rural vs. urban (<100,000 and 100,000+) and east vs. west. All rural counties with four or more Family Care Homes are eligible for selection in the sample design. Twelve rural counties from the 32 in the east region will be randomly selected and 12 rural counties from the 36 eligible counties in the west region. In each of the randomly selected 24 counties, two Family Care Homes will be randomly selected for enrollment in the study.

From the 15 urban counties, those with ten or more family care homes are eligible for selection (10 counties). All the counties with more than 200,000 inhabitants will be forced in the sample, while three urban counties will be randomly selected from the remaining five counties. Six Family Care Homes will be randomly selected from each of these urban counties. Family Care Homes in the East Rural, West Rural and Urban regions will be randomly assigned to the control group, the experimental group with a three month interval, six month interval and nine month interval. See Table 4 for a display of the Family Care Home random assignment.

TABLE 3
COUNTIES WITH MORE THAN 5 HOMES AND URBAN COUNTIES WITH MORE THAN 100,000 INHABITANTS

COUNTY	# FACILITIES IN COUNTY	URBAN/RURAL*		REGION	
ALAMANCE	25	X		WEST	103,229
ALEXANDER	2		X	WEST	26,814
ALLEGHANY	1		X	WEST	9,722
BEAUFORT	5		X	EAST	43,396
BERTIE	3		X	EAST	21,251
BLADEN	8		X	EAST	30,826
BRUNSWICK	1		X	EAST	47,797
BUNCOMBE	46	X		WEST	170,004
BURKE	13		X	WEST	75,900
CABARRUS	6		X	WEST	92,844
CALDWELL	5		X	WEST	70,146
CARTERET	3		X	EAST	50,414
CASWELL1	3		X	WEST	22,489
CATAWBA	2	X		WEST	114,143
CHATHAM	2		X	EAST	36,015
CHEROKEE	3		X	WEST	20,363
CLEVELAND	17		X	WEST	86,216
COLUMBUS	2		X	EAST	52,292
CUMBERLAND	11	X		EAST	254,943
CURRITUCK	2		X	EAST	11,089
DAVIDSON	4	X		WEST	119,094
DUPLIN	5		X	EAST	41,685
DURHAM	39	X		EAST	165,839
EDGECOMBE	2		X	EAST	59,071
FORSYTH	21	X		WEST	260,853
FRANKLIN	9		X	EAST	34,173
GASTON	3	X		WEST	171,784
GATES	1		X	EAST	8,875
GRANVILLE	4		X	EAST	37,696
GREENE	1		X	EAST	16,117
GUILFORD	28	X		WEST	329,862
HALIFAX	2		X	EAST	50,030
HARNETT	4		X	EAST	64,009
HAYWOOD	4		X	WEST	48,469
HENDERSON	4		X	WEST	67,222
HERTFORD	10		X	EAST	24,046
IREDELL	3		X	WEST	88,429
JOHNSTON	11		X	EAST	78,191
LEE	3		X	EAST	36,718
LENOIR	5		X	EAST	60,220
LINCOLN	1		X	WEST	42,372
MADISON	5		X	WEST	16,827
MARTIN	5		X	EAST	26,719
MCDOWELL	9		X	WEST	36,220
MECKLENBURG	17	X		WEST	453,107
MITCHELL	6		X	WEST	14,428
MONTGOMERY	5		X	WEST	23,852
MOORE	4		X	WEST	56,009
NASH	5		X	EAST	71,241
NEW HANOVER	10	X		EAST	114,656
NORTHHAMPTON	2		X	EAST	22,497
ONSLOW	7	X		EAST	125,134
ORANGE	11		X	EAST	84,729

TABLE 3 (continued)
COUNTIES WITH MORE THAN 5 HOMES AND URBAN COUNTIES WITH MORE THAN 100,000 INHABITANTS

COUNTY	# FACILITIES IN COUNTY	URBAN/RURAL*		REGION	
PASTQUOTANK	1		X	EAST	28,462
PENDER	5		X	EAST	25,199
PERQUIMANS	4		X	EAST	12,534
PERSON	2		X	WEST	30,648
PITT	16		X	WEST	97,406
POLK	4		X	WEST	14,486
RANDOLPH	12		X	WEST	99,070
RICHMOND	5		X	WEST	46,227
ROBESON	11	X		EAST	106,094
ROCKINGHAM	25		X	WEST	85,516
ROWAN	7	X		EAST	104,523
RUTHERFORD	27		X	WEST	55,880
SAMPSON	5		X	EAST	50,321
SCOTLAND	6		X	WEST	33,735
STANLY	1		X	WEST	50,276
STOKES	1		X	WEST	35,610
TRANSYLVANIA	2		X	WEST	23,417
TYRRELL	1		X	EAST	3,575
UNION	10		X	WEST	79,832
VANCE	5		X	EAST	38,740
WAKE	34	X		EAST	366,004
WARREN	4		X	EAST	16,452
WASHINGTON	2		X	EAST	14,801
WATAUGA	2		X	WEST	34,497
WAYNE	5		X	EAST	97,410
WILKES	1		X	EAST	60,727
WILSON	6		X	EAST	64,564
YADKIN	1		X	WEST	29,643
YANCEY	1		X	WEST	15,843

* URBAN 100,000 OR MORE INHABITANTS
RURAL LESS THAN 100,000 INHABITANTS

TABLE 4
Distribution of Family Care Homes by Characteristics Over the Study Groups

	control XY	3 months AB	6 months CD	9 months EF
East Rural	6	6	6	6
West Rural	6	6	6	6
Urban	12	12	12	12
	24	24	24	24

References

ASCP, Detecting and Counting Medication Errors in Long-Term Care Facilities, A Self-Instructional Surveyor Training Manual. American Society of Consultant Pharmacies, Arlington, VA, 1984.

Avorn, J., Dreyer, P., Connelley, K., Soumerai, S.B., Use of Psychoactive medication and the quality of care in Rest Homes. New England Journal of Medicine, 1989.

Barker KN, Pearson RE, Hepler CD, Smith WE, Pappas CA, Effect of an automated bedside dispensing machine on medication errors. Am J Hosp Pharm 1984; 41:1352—1356.

Barker KN, et al. Medication errors in nursing homes and small hospitals. Am J Hosp Pharm 1982; 39:987—991.

Barker KN. The effects of an experimental medication system on medication errors and costs: I. Introduction and error study. Am J Hosp Pharm 1969; 26:324—333.

Barker KN. The problems of detecting medication errors in hospitals. Am J Hosp Pharm 1962; 19:360—369.

Bradshaw BR, Vonderhaar WP, Keeney VT, Tyler LS, Harris S. Community-based residential care for the minimally impaired elderly: a survey analysis. J Am Geriatr Soc 1976; 24:423—429.

Budden F. Adverse reactions in long-term care facility residents. J Am Geriatr Soc 1985; 33:449—450.

Dyer CC, Oles KS, Davis SW. The role of the pharmacist in a geriatric nursing home: a literature review. Drug Intell Clin Pharm 1984; 18:428—433.

Eckel FM, Crawley HK, III. A study of pharmacy services in North Carolina rest homes. J Am Pharm Assoc 1971; 11:387—390.

Ferry ME, Lamy PP, Becker LA. Physicians' knowledge of prescribing for the elderly. Presented at Northeast Gerontological Society Meeting, April, 1984.

Gibson RM, Mueller MS, Fisher CR. Age difference in health care spending, fiscal year 1976. Soc Security Bull 1977; 40:3.

Godwin-Austen RG, Frears CC, Bergmann S. Incidence of side effects from Levodopa during the introduction of treatment. Br Med J 1971; 1:267—268.

Goetz CG, Weiner WJ, Nausieda PA, Klawans HL. Tardive Dyskinesia: Pharmacology and Clinical Implications. Clinical Neuropharmacology 1985; 5:3—22.

Hartzema AG, Godbout NP, Lee SD, Konrad TR, Eckel FM. Evaluation of an educational intervention to reduce medication administration errors in domiciliary care facilities. Unpublished manuscript, 1986.

Hartzema AG, Konrad TR. Drug prescribing frequencies in nursing facilities and homes for the aged: patient care implications. Am J Public Health (in review).

Hartzema AG, McDermott JM, Konrad TR. Drug prescribing frequencies in extended care facilities. American Pharmaceutical Association Congress, San Antonio, Texas, February 19, 1985.

Health Care Financing Administration. State Operations Manual. Provider Certification, Publication 7, Transmittal No. 174, March, 1985.

Hollister LE. General principles of treating the elderly with drugs. In: LF Jarvik, DJ Breenblatt, D Harman. Clinical Pharmacology and the Aged Patient, Raven Press, New York, 1981.

Holloway DA. Drug problems in the geriatric patient. Drug Intell Clin Pharm 1974; 8:632—642.

Jameson, B. Oral communication. 1988.

Jenke M. Tardive dyskinesia: special risk in the elderly. J Am Geriatr Soc 1983; 31:71—73.

Kalchthaler T, Coccara E, Lichtiger S. Incidence of polypharmacy in long-term care facility. J Am Geriatr Soc 1977; 25:308—313.

Kane RL, Kane RA, Arnold SB. Prevention and the elderly: risk factors. Health Serv Res 1985; 19:945—1006.

Lamy PP. Special features of geriatric prescribing. Geriatrics 1981; 36:42—52.

Larson EB, Kukulh WA, Buchner D, Reifler BV. Adverse drug reactions associated with global impairment in elderly persons. Ann Intern Med 1987; 107:169—173.

Law R, Chalmers C. Medicines and elderly people: a general practice survey. Brit Med J 1976; 1:565—568.

Lee S. The organization of pharmaceutical services in domiciliary care facilities. Unpublished masters thesis, University of North Carolina at Chapel Hill, 1987.

Meier DE, Myers WM, Swenson R, Bennet WM. Indomethacin-associated hyperkalemia in the elderly. J Am Geriatr Soc 1983; 31:371—373.

Mellody JF, White JG. Service delivery assessment of boarding homes, Philadelphia: U.S. DHEW, Office of the Principal Regional Official, DHEW Region III (May 1979).

Mor V, Sherwood S. A national study of residential care for the aged. The Gerontologist 1986; 26:405—417.

Morse ML, Illig NA, Barker K. Drug errors: counting and controlling. In: Pharmacy Issues in Long-Term Care. National Foundation for Long-Term Care, Washington, DC, 1982.

Nelson GM. Domiciliary care for the aged: implications for the states. Unpublished manuscript, University of North Carolina at Chapel Hill, 1988.

N.C. Office of State Budget and Management, January, 1988.

Palumbo FB, Magaziner JS, Tenny JH, Goren LM, Warren JW. Recruitment of long term care facilities for research. JAGS 1987; 35:154—158.

Parkin DM, Henney CR, Quick J, et al. Deviation from prescribed drug treatment after discharge from hospital. Brit Med J 1976; 2:1976.

Ray WA, Blazer DJ, Schaffner W, Federspiel CF. Reducing antipsychotic drug prescribing for nursing home residents: a controlled trial of the effect of an educational visit. Am J Public Health 1987; 77:1448—1450.

Scanlon W and Sulvetta M. The supply of institutional long-term care: descriptive analysis of its growth and current state. The Urban Institute, Work Paper, 1466-13 (rev.), April, 1983.

Schwartz D, Wang M, Zeitz L, et al. Medication errors made by elderly chronically ill patients. Am J Public Health 1962; 52:2018.

Sherwood S, Morris JN. The Pennsylvania domiciliary care experiment, I. Impact on quality of life. Am J Public Health 1983; 73:646—653.

Sloane P, Lekan-Rutledge D, Gilchrist P. Telephone contacts in the decision-making process. J Geront Nursing 1986; 12:35—39.

Stone R, Newcomer RF. The state role in board and care housing. In: Long Term Care of the Elderly Ed. C. Harrington, et al., Sage Publications, Beverly Hills, CA, 1985.

Vener AM, Krupka LR, Climo JJ. Drug usage and health characteristics in non-institutional retired persons. Am Geriatr Soc 1979; 27(2):83—90.

White TR, Anastasio GD. PRN medications in rest home setting. North Carolina Medical Journal 1988; 42:43—45/

THE UNIVERSITY OF NORTH CAROLINA
AT
CHAPEL HILL

Committee on the Protection
of the Rights of Human Subjects
The School of Medicine
(919) 966-1344

The University of North Carolina at Chapel Hill
CB# 7000, MacNider Building
Chapel Hill, N.C. 27599-7000

TO: Abraham G. Hartzema, Ph.D., M.S.P.H.
School of Pharmacy
CB #7360 Beard Hall
Campus

FROM: [redacted], Chairman of the Committee on the Protection of the Rights of Human Subjects

DATE: June 15, 1989

SUBJECT: Research Application Review

Number: 89-PHARM-79

Title: Reducing Medication Administration Errors in Family Care Homes Through a Program of On-Site Educational and Technical Assistance

This research proposal has been considered by the Committee and it has been approved until June 15, 1990.

NOTE:

(1) This Committee complies with the requirements found in Part 56 of the 21 Code of Federal Regulations and Part 46 of the 45 Code of Federal Regulations. Assurance Number: M-1390; IRB Number: 01NR.

(2) Re-review of this proposal is necessary if:
(a) Any significant alterations or additions to the proposal are to made
(b) You wish to continue beyond the above date

(3) It is required that all signed consent forms be retained on file.

(4) A copy of the approved consent form is enclosed. This is the only consent form approved for this study.

Chairman of the Committee

MANAGEMENT OF DRUG THERAPY IN REST HOMES

Principal Researcher:* Abraham G. Hartzema, Ph.D.
Telephone: (919) 962-0080

I understand that the study involves research in the management of drug therapy in family care homes. The purpose of the research is to develop and implement procedures and record keeping practices to make drug therapy more efficient and effective for the elderly residents in my home.

I consent that one of the researchers will observe me while I am administering medications at two separate occasions on the first day and on one occasion at a later time interval. I agree that notes will be taken during these observations. For participating in the program I will receive Continuing Education credits. I understand that no risk or discomfort for the resident or others involved is expected although some risks may be unforseeable.

I understand that the observations made are confidential, and that this information will not be shared with owners of the home, administrators or supervisors nor with any oversight agencies. This information will only be used combined with similar information obtained from other homes.

I understand that I am free to refuse to participate or to withdraw from the research activity at any time without penalty.

I understand that this project has been approved by the Committee on the Protection of the Rights of Human Subjects at the University of North Carolina, Chapel Hill. If I believe that there is any infringement upon my rights, I may contact the Chairman of the Committee, [redacted] at 919-966-1344.

I have the opportunity to ask, and have answered, all my questions about this research.

I will contact the Principal Researcher, Dr. Abraham G. Hartzema, at 919-962-0080 if there are any further questions about the study.

I have read and understand the consequences of the study.

______________________________	______________________________
Administrator or Supervisor in Charge/date	Witness/date

* The term "principal researcher" is used here instead of "principal investigator" because it became evident from previous research endeavors in this care segment that the latter term was reserved for state investigators investigating violations of licensing requirements.

CONSENT FORM

APPROVED ON:

6-15-89 JCH/bsb

C. EXAMPLE OF AN APPROVED BUT UNFUNDED PROJECT: SMOKING PREVENTION IN BLACK CHILDREN

APPLICATION FOR A RESEARCH OR CLINICAL INVESTIGATION GRANT

(Please read carefully the attached "Policies on Research and Clinical Investigation Grants" and the instructions on all pages of this form, before completing this application.)

TO: [redacted]

Date July 25, 1988

Application is hereby made for a grant* in the amount of $107,729.42 **for the period from** July 1989 **to** July 1991 **inclusive.**

Title Smoking Prevention in Black Children

FROM:

Name of Investigator **Signature**

Associate Professor
Title **Telephone No.**

College of Pharmacy & Pharmaceutical Sciences
Department **Division of Institution**

Tallahassee, Florida 32307
Street and Number **City, State, Zip Code**

Florida Agricultural & Mechanical University
Official Name of Institution

Tallahassee, Florida 32307
Street and Number **City, State, Zip Code**

Name of Official Authorized to Sign for Above Institution Signature

University President
Title

*It is understood that each applicant, by the act of applying for a grant, agrees. If the grant is made, to abide by the Society's POLICIES ON RESEARCH AND CLINICAL INVESTIGATION GRANTS.

Florida A&M University
Division of Sponsored Research
Checks to be Made Payable to

(Re; Research title)
Street and Number

Tallahassee, Florida 32307
City, State, Zip Code

CERTIFICATE FOR PROTECTION OF HUMAN SUBJECTS AND CONTAINMENT FOR RECOMBINANT DNA RESEARCH

1. PRINCIPAL INVESTIGATOR ______________________________

2. TITLE OF PROPOSAL Smoking Prevention in Black Children

XXX

3. RESEARCH INVOLVING HUMAN SUBJECTS

Safeguarding the rights and welfare of human subjects involved in activities supported by grants from ████████ is the responsibility of the institution which receives the funds awarded. In order to provide for the adequate discharge of this institutional responsibility, no grant for an activity involving human subjects shall be made unless the application for such support has been reviewed and approved by an appropriate institutional committee.

CHECK ONE OF THE FOLLOWING STATEMENTS AS APPLICABLE:

☐ A. This application does not propose any activities that would involve human beings as research subjects.

☐ B. This is to CERTIFY that this application which does propose activities involving human subjects has been reviewed and approved by our institutional committee on the date of ________ in accordance with current DHHS policy. (The review date should be recent; certification is invalid if review date would precede award date by more than one year.)

XXX Exemption #2&3 according to NIH guidelines.

---

4. RESEARCH INVOLVING RECOMBINANT DNA MOLECULES

The physical and biological containment necessary for research involving recombinant DNA molecules (as defined in the current "NIH Guidelines for Research Involving Recombinant DNA Molecules") supported by grants from ████████ ████ is the responsibility of the institution which receives the funds awarded. In order to provide for the adequate discharge of this institutional responsibility, no grant for an activity involving recombinant DNA molecules shall be made unless the application for such support has been reviewed and approved by an appropriate institutional committee.

CHECK ONE OF THE FOLLOWING STATEMENTS AS APPLICABLE:

☐ A. This application does not propose any activities that would involve recombinant DNA molecules, as defined by current NIH guidelines.

☐ B. This is to CERTIFY that this application which does propose activities involving recombinant DNA molecules has been reviewed and approved by our institutional committee on the date of ________ in accordance with current NIH guidelines. (The review date should be recent; certification is invalid if review date would precede award date by more than one year.)

---

5. Signature of Institutional Official Authorized to Sign Proposals **Date:**

University President 904-599-3225

6. Title **Telephone No. (Code No. Extension)**

CERTIFICATION FOR CARE AND TREATMENT OF LABORATORY ANIMALS

1. **PRINCIPAL INVESTIGATOR** ______________________________

2. **TITLE OF PROPOSAL** Smoking Prevention in Black Children

3. **RESEARCH INVOLVING LABORATORY ANIMALS**

The proper care and humane treatment of laboratory animals involved in activities supported by grants from [redacted] is the responsibility of the institution which receives the funds awarded. In order to provide for the adequate discharge of this institutional responsibility, no grant for an activity involving laboratory animals shall be made unless the application for such support has been reviewed and approved by an appropriate institutional committee.

CHECK ONE OF THE FOLLOWING STATEMENTS AS APPLICABLE:

[X] **A. This application does not propose any activities that would involve laboratory animals.**

[ ] **B. This is to CERTIFY that this application which does propose activities involving laboratory animals has been reviewed and approved by our institutional committee in accordance with current NIH policy.**

4. **LIST SPECIES OF LABORATORY ANIMALS TO BE USED.**

NONE

____

5. **Signature of Institutional Official Authorized to Sign Proposals** **Date:**

University President — 904-599-3225

6. **Title** — **Telephone No. (Code No. Extension)**

APPLICATION FOR A RESEARCH OR CLINICAL INVESTIGATION GRANT

1. BUDGET PROPOSED* (Refer to Instruction Sheet)

	Title	%Time	1st Year	2nd Year
A) PERSONNEL				
(Itemize all positions; indicate per cent of time, salary and names of personnel, if selected)				
*Dr.	Princip. Inv.	25%	12,500	13,250
*Dr.	Co-Investi.	20%	7,600	8,056
*Mr.	Counselor	10%	2,500	2,650
Vacancy $5.00/hour	Stud. Asst(s)	50%	5,600	5,600
Grade School Teachers (To be identified)		10%		
			$28,200	$29,556
Retirement or other security benefits (Total)	*34%*		5,369	5,687
	CATEGORY TOTAL		33,569	35,243
B) PERMANENT EQUIPMENT *(ITEMIZE)*				
Audiovisual Equipment			2,000	–
	CATEGORY TOTAL		2,000	-0-
C) CONSUMABLE SUPPLIES *(GROUP IN MAJOR CATEGORIES)*				
Questionnaire Printing (Including Pretest/Post-test, Posters & Phamphlets			2,100	1,900
Audiovisual Cassettes			900	900
Drawings, Coloring Paper, & Crayola			2,200	2,200
Computer Statistician Analysis & Software			1,000	1,000
Equipment Maintenance			1,000	1,000
Paper for (Computer) Printer			500	500
	CATEGORY TOTAL		7,700	7,500
D) MISCELLANEOUS *(LIST SPECIFIC AMOUNT FOR EACH ITEM)*				
Travel to Professional Meetings in Relation to the Research such as Presentations			1,500	1,500
	CATEGORY TOTAL		1,500	1,500
	SUBTOTAL		44,769	44,243
E) INDIRECT COSTS 25% Direct Cost minus equipment			11,192	11,061
([redacted] Policy)	**GRAND TOTAL**		~~55,961~~	~~55,304~~
			$56,000	$55,000

***See POLICIES ON RESEARCH AND CLINICAL INVESTIGATION GRANTS**

If more space is needed prepare budget using above format.

NO. 0059.02 - 1/82

APPLICATION FOR A RESEARCH OR CLINICAL INVESTIGATION GRANT

SUMMARY OF RESEARCH PROPOSED

Name and Official Title of Principal Investigator

College of Pharmacy and Pharmaceutical Sciences, Florida A&M University

Name and Address of Applicant Organization

Smoking Prevention in Black Children

Title of Project

Smoking Prevention in Black Children

Use this space to summarize concisely your proposed research. Outline objectives and methods. Underscore the Key words (not to exceed 10) in your abstract.

SMOKING PREVENTION IN BLACK CHILDREN

Objectives: The long term objectives of this research project are smoking prevention and subsequent decrease in lung cancer among black populations. The specific aims are to: test the effectiveness of achieving early smoking prevention in black children through program education on (a) the hazards of smoking; (b) how to deal with peer pressures; (c) deal with some of the myths associated with smoking; (d) encourage the children to take home with them what they learned from the project, and help influence members of their family who smoke; and (e) study the effect of educational strategies in black children in smoking prevention. Location: Florida A&M University Elementary, Tallahassee, Florida. (Historically Black Elementary School, which is a laboratory demonstration school for the university). Method: To achieve the objective and specific aims, audiovisuals, pamphlets, and puppets will be used. The program will be developed and used for black 4th graders. It is believed that children try smoking at an early age and thus, it is the best time to teach them not to smoke early. The 4th graders will be followed-up for four years with re-inforcement smoking prevention and cessation programs to the end of the 7th grade. At the end of each grade, the children will be evaluated as to their knowledge of the effects of smoking using pre-test and post-test questionnaires. At the end of the program (4th year), questionnaires will be used in the overall program evaluation to determine the number of children who have attempted and/or started smoking and effects if any/ or smoking family members. Evaluation: For statistical evaluation, for each grade, the children involved in the study will constitute the study group, while those in the same grade, in a comparable black elementary school, will serve as the control group. Also, having the pre-test and post-test date from the study group will enable the children to act as their own individual control during the annual analysis.

Leave Blank Below This Line

No. 0059.02 - 1/82

BUDGET JUSTIFICATIONS

The principal investigator and research associate will be involved in several phases of the development, presentation and evaluation of the project with the elementary school children during the entire project period.

Personnel

The principal investigator will be responsible for the overall project administration, implementation and evaluation. Both the research associate and student assistants will assist the principal investigator in these various areas. The principal investigator is experienced in questionnaire design, evaluation and statistical analysis. [redacted] is an assistant professor with the University elementary school where the study will take place. He is an experienced child education psychologist whose background will enhance the program effectiveness with the children. Augustine Okolo brings a vast experience with personal computers and child education to the project. Augustine is a research associate with the College of Pharmacy who handles most of the personal computer needs of the College. He is also experienced in the area of child education and has worked with both the Texas School system and the Florida School system. His contribution to this project cannot be over emphasized. He has also taught elementary school children including Black children, for several years. Augustine will train and supervise students on all data entry into the personal computer. He will also be responsible for all computer related tasks in relation to the project.

EQUIPMENT:

The audiovisual equipment and cassettes are needed to show films and pertinent information to the children in the study group. This is extremely important for the project. The IBM personal computer and accessories are needed for data storage, analysis and reporting of research results. It is important that the project has a personal non-shared computer and audiovisual equipment to evaluate the project to be more efficient and meet deadlines in the areas of data storage, analysis and reporting. Shared equipment will result in problems of accessibility and missing deadlines. It is important to note that both the nature and duration of the study will result in collection of large amounts of data from ongoing pretest and post-test.

SUPPLIES:

The questionnaire pretest and post-test are vital components of the research study. Without these effective program evaluation will be extremely difficult if not impossible. The pamphlets and posters are part of the project tools for implementation.

TRAVEL:

Both the principal investigator and selected students will travel to professional meetings to present the research results. Anticipated meetings include the American Pharmaceutical Association (APhA) and American Public Health Association (APHA). These presentations at professional meetings will give the students an excellent exposure to the professional and scientific community.

OTHER EXPENSES:

The equipment maintenance expense is necessary to assure that equipments are maintained in good working condition.

The student assistants will be involved in both the implementation and administrative portion of the project (typing, filing). They will also be involved in the data analysis and evaluation, i.e., entering data into the computer for analysis. Student presentations will be highly encouraged during research paper presentations to the health professional community in relation to the project, especially at professional association meetings. The students will consist of an undergraduate student and two doctor of pharmacy students. The project will serve as an excellent educational goal for the students as well as get them involved in research activities. The students will enhance the overall productivity of the project.

I. WORK ACCOMPLISHED BY P.I. AND/OR OTHERS

The Principal Investigator (P.I.) hs been involved in various research activities in the area of health, such as blood pressure medication compliance, drug substitution study, international health in developing nations (Africa), medical applications in computer technology, study on early warning signs of cancer, and attitudes of blacks towards cancer. Some of these research activities have resulted in journal publications, which are listed under publications in the professional qualifications and research section of this grant application. The attitudinal cancer research study among blacks is in the final stages of data analysis; as a result the results of that study are not available.

Several studies have been reported in the literature in the area of smoking prevention & intervention, and smoking cessation methods. Some of these studies involve employer sponsored and work site programs. Such studies include that by Fielding, J.E., reported in Annual Review of Public Health 5, 1984 and that of Kapin, K. in Corporate Fitness and Recreation, October-November 1984. Some of the research studies and projects involving children (ages 10 and above) were done outside the United States, such as Murray et al. study in Journal of Epidemiology and Community Health, September 1984 which was done in Northern England. The study by Tell G.S. et al. which is reported in Preventive Medicine, May 1984 was done in Norway.

Over 100 (one hundred) different published studies and projects in the area of smoking prevention & intervention; and smoking cessation including adverse effects of smoking was reviewed in relation to this grant application and proposal. However, none of the studies focused on black children. Even the studies that did focus on non-black children concentrated their attention more on children 10 years and above. This study proposes to target black children at an earlier age and then follow them through with re-enforcement, to reduce the incidence of smoking and subsequently lung cancer associated with smoking in blacks. The black population has not received significant attention in research studies involving cancer/smoking related research.

AIM, METHOD & SIGNIFICANCE OF THE STUDY

2. Specific Aims

The specific objectives of this research project are to prevent smoking in black children and subsequently decrease possible lung cancer. Smoking often starts at an early age; therefore, for smoking prevention programs among blacks to be effective, such programs should be started early in grade school. Through the project, black grade school children will be aware of the effects and consequences of smoking and how to deal with peer pressure to smoke, and one can effectively and efficiently reduce the incidence of smoking among blacks. Children can also be effective in helping other family members stop smoking through their interaction due to their exposure to the smoking prevention and cessation program. Thus, the program will try to discourage children from smoking and help them to grow up as non-smokers. The program will also try to influence adult family members who are smokers to quit smoking through their childrens' influence. The specific aims are: To test the effectiveness of the educational program in achieving early smoking prevention in black elementary school children through program education in (1) the hazards of smoking, to achieve early smoking prevention in these black children; (2) how to deal with peer pressures to smoke; (3) deal with some of the myths associated with smoking such as, smoking makes one sophisticated or grown up etc., and (4) encouraging the children to help influence family members who smoke to stop. Hypothesis: (1) The study group will be better informed about the dangers of smoking compared to the control group; (2) the number of smokers in the study group will be less than that of the control group at the end of the 4-year program period. Student Participation: Trained students will be involved in all phases of the project including presentation of some information to the children, data collection and storage (entering information in the computer for analysis), and project reports. Student study presentations, where appropriate, will be encouraged during health professional organizational meetings, such as the American Public Health Association (APHA) and the American Pharmaceutical Association (APhA). The students will consist of one undergraduate student and two doctor of pharmacy students. The project will serve as an opportunity for (1) student research and (2) student research findings presentation to the professional and scientific community during project presentation in a professional meeting.

Significance

The significance of this research project is that, through early smoking prevention and cessation in black children, the incidence of smoking related morbidity and mortality among blacks will be reduced. Cigarette smoking is still a major preventable cause of illness and death.[1-2] The incidence of lung cancer among blacks is rising, yet the level of knowledge of this disease and its prevention is low.[3] Blacks are more likely to get cancer and die from it than whites.[4] The Department of Health and Human Services "Smoking and Health", a national status report to Congress in 1986, indicated that there are more black (males and females) smokers than whites. Thus, blacks are a high risk group in the area of smoking and lung cancer. The high incidence of smoking with resultant lung cancer in blacks may be a result of the lack of educational programs for blacks, as well as initiating preventive programs late when smoking has already been initiated.

School enrollment among blacks has increased significantly over the last decade.[5] Black youth are the future of the black population. Providing black children with smoking prevention and cessation programs in their early years can help to improve the future health of the black population. By targeting black elementary children in historically black schools, for example, their incidence of smoking and subsequent lung cancer can be reduced.

Several smoking prevention and cessation strategies have been developed, such as behavioral techniques, television, etc.; however, none of these programs have targeted the black population.[6-12] Little or no smoking prevention and cessation study has been conducted for black school children in the 4th or lower grades. However, some studies have been done in some predominantly white and multi-racial high schools. A longitudinal study via personal household interviews done by Kandel and Logan indicated that the risk of smoking initiation began to increase at age 9, and peaked at about age 16.[16] A survey of Illinois school students (grades 7 through 12, ages 11 to 20) indicated that the majority of the students believe that smoking is a real health problem and not socially acceptable.[17] Therefore, while information and studies have concentrated on children 9 and above, very few studies have been done with younger children. Researchers agree that a smoking prevention program should start with younger children,[16,28,30] a view also held by Oei et al.[18] as a result of their study with nine year olds. They concluded that smoking prevention programs for children should start at a younger age,[19] prior to age 9.[18,30] Studies have been done which emphasized the importance of school-based anti-smoking programs as a means of preventing the onset of cigarette smoking. In these studies the study groups were better informed about smoking and are less likely to be smokers than those without the program.[20-29]

Several factors influence smoking in adolescents, such as smoking parents, smoking siblings, peers, smoking approval from parents, and self esteem.[31-32] Murray[32] indicated that single parent families are more likely to smoke regardless of parental smoking habits.[32] A biracial study by Croft et al. with children (ages 8 to 17) concluded that smokers are not only most likely to have friends and family members who smoke but also are most likely to have purchased their first cigarette and believe smoking is pleasurable for themselves and others.[34]

Noting that (1) more blacks than whites smoke, (2) black adolescents especially males have the highest prevalence of tobacco use and the lowest knowledge about its dangers,[35] (3) children can start experimenting with cigarettes as early as 9 or younger,[30] (4) more of the individuals who start smoking early tend to be smokers (Smoking Health National Status Report) makes the black population more vulnerable and a high risk group to the dangers of smoking. Therefore, this research project will be longitudinal and focus on black children in the fourth grade until the end of their seventh grade year.

The project will inform them of the dangers of smoking, influence them not to smoke, and potentially influence family members or friends who smoke. Therefore, the program will start prior to fourth grade and pass through the sixth grade, which may be the major period of smoking onset.[28] The program will end after their seventh grade year. Therefore, the program will result in early smoking prevention in black children, who will grow up as non-smoking adults, thus reducing all the health risks associated with smoking in blacks. This makes the project unique because of the population in black elementary school children.

Experimental Design Methods and Evaluation

The smoking prevention and cessation program will involve the use of audiovisuals, pamphlets, drawings, puppet shows, coloring books and posters. These materials will be developed specifically for black fourth graders at the historically black Florida A&M University Elementary School. The research project will emphasize smoking prevention. However, the children will also be provided with information for friends and relatives who are smokers. Some of these materials will be obtained from the American Cancer Society.

The following programs and materials may be modified and used in overall program development:

(1) "Smoking Deserves a Smart Answer School Health Program" from the American Lung Association of Broward - Glades - Hendry, Inc. The Program is designed for fifth and sixth graders and addresses the following:
  a. Health effects of smoking
  b. Social and behavioral motivations related to smoking
  c. Skills for identifying and resisting peer pressure
  d. Effects of cigarette advertising
  The program contains puzzles, posters, parent information, and resource folders of classroom activities.

(2) "Smoking Intervention and Prevention Programme". From the American Lung Association of Florida, Spaceport Branch. The program has different films developed for students in grades K—4, 5—8 and 9—12. Programs are illustrated with films, smoking manikin, coloring books, slides, booklets, posters, signs, buttons, and quizzes.

(3) Smoking and Health Teaching Unit for grades K—3, 4—6 and high school. From: Brevard County Interagency Council on Smoking and Health, Rockledge, Florida.

(4) "Octopuff in Kumquat" program of American Lung Association for elementary school children.

(5) "A Manual on Smoking and Children". Edited by Wake, R., McAlister, A., Nostbakken, D., includes various approaches for controlling smoking in children including:
   a. preventive education
   b. dissemination of factual information
   c. peer programs
   d. skills training
(6) "Know Your Body Program". It is essential that some of these programs be modified to be more specific for black children and their families. The importance of using programs which are specific to a cultural group was emphasized by Dignau et al.[35]

The study group will consist of black children in the fourth through seventh grade year 1 to 4 of the study in the Florida A&M University Elementary School, while the control group will consist of black fourth to seventh graders in a comparable school. Qualitative and quantitative information using pre-test/post-test control group design with a non-equivalent, control group to assess (1) the children's knowledge of health consequences of smoking, (2) resistance to persuasion skills, (3) cigarette smoking, (4) collect demographic information, (5) determine experimentation with cigarettes, and (6) determine the number of children with family members and friends who smoke. Their collected information will be analyzed annually starting with the end of the first year to determine program effectiveness. The same pre- and post-test questionnaire will be administered to both the study group and control group. This will aid in identifying and analyzing differences between the study group and the control group.

The program will consist of a combination, and modification, of some of the presently available programs for fourth graders, as has already been indicated, thereby tailoring the program to meet the needs of the children. The program will be conducted for one hour per week, during the school year. The project, although structured to be flexible enough to accommodate any changes, will improve the effectiveness of the program to achieve overall goals.

Similar approaches will be used for the same children (study and control) while in fifth, sixth, and seventh grades for reinforcement. However, the program materials will be upgraded and different for each subsequent grade so as to be suitable for that particular age group. Thus a pre-test questionnaire will be administered and evaluated. The overall program will be evaluated at the end of the project (4 years).

The pre-test and post information collected during the annual evaluation will be entered into the computer and analyzed. One of the most important aspects of data analysis is the selection of appropriate statistics for the data results to be interpretable or meaningful. Different types of data will be generated from the questionnaire, the pre-test and post-tests which will be given to the children as a part of the program evaluation. Cross tabulation (a joint frequency distribution according to two or more classified variables) will be used to illustrate relationships between variables through the numerical tabular presentation of data in frequency or percentage form. Statistical tests using the chi-square and t-test statistics at .05 level of significance will be used to analyze and evaluate collected data and, at the end of the project period, to determine data significance and achievement of objectives. These test statistics will be used also to test the hypothesis that the study group will be better informed about smoking and less likely to be smokers than the control group. Both the chi-square and the t-test statistics are appropriate for this type of project. Also the relationship between demographic variables will be evaluated.

The project will start with the black fourth graders, and the children will be followed through the end of seventh grade for reinforcement and program enhancement because several studies (as had been indicated) concluded that (1) blacks are less knowledgeable about the health consequences of smoking, (2) smoking prevention programs should start prior to sixth grade, (3) preventive rather than intervention programs should be emphasized,[30] (4) programs be more specific to cultural groups,[35] and (5) school-based programs are beneficial in smoking prevention. It is necessary for the program to run for four years to increase the program's effectiveness and achieve designed overall goals. The four year period will cover the 6th grade which is a major period of smoking onset.[28]

Potential Problems and Remedies:

School children in the study or control group who move out of the study area will be excluded from further evaluation. Children who move from the study group into the school for control group will not participate as a control group because they have already been exposed to the project as a study group; however, any influence that children might have on peers will be noted. If, however, a child moves from the control group into the study group, the child can participate in the study group and appropriate adjustments made during evaluation.

Overall Project Outline for Year 1 to 4

(Fourth through seventh grade-same students)

	Study Group	Control Group
Beginning of each grade	Pre-test questionnaire	Pre-test questionnaire
Collected data entered into computer	Program	No Program
End of Program for each grade	Post-test	Post-test

Post-test data entered into computer and program evaluation/ analysis is done at the end of the program for each grade.

Overall program evaluation and analysis will be performed after the four year period.

3. Facilities Available

The University is a historically black public (State) university serving the needs of the black community. It's adjacent to the University elementary school where the elementary school children will be involved in the project. The College of Pharmacy has a very conducive environment for this project, with several faculty members with research experience, type-setting equipment, etc. Office will be designated for the project when funded, furnished with desks, chairs, typewriters, etc., within walking distance to Florida A&M University Elementary School.

Institutional Qualifications — Grants and Contracts

Florida A&M University, which is state public institution, attracts over 12 million dollars in direct costs from contracts and grants per fiscal year. Although FAMU has the smallest student population, it ranks fourth among the nine universities of the Florida State University System in securing sponsored research funds. The faculty of the College of Pharmacy - FAMU attracts the largest proportion of these research funds university-wide. The College of Pharmacy faculty has been successful in securing funds from the National Aeronautics and Space Administration, March of Dimes Birth Defects Foundation, Agency for International Development, National Institutes of Health and Department of Education. Florida A&M University in 1983—84 joined the approximately two (2) other HBC's in eligibility of the number of research support from the National Institutes of Health because of the number of R01, S01, S03 and MBRS awards generated by the faculty of the College of Pharmacy.

Since 1980, the College of Pharmacy faculty has published over 100 scientific articles in refereed journals including science and official publications of different professional societies. Many of these scientific publications received national attention and were the subject of various news and press releases. Faculty also presented over 120 papers at national and international non-MBRS meetings and supervised student presentations at MBRS meetings which average approximately 30 presentations per year.

The College of Pharmacy and Pharmaceutical Sciences is highly qualified to administer and carry out research activities such as this one. The College of Pharmacy has successfully secured funding for research and training activities from the State, national as well as international level. The College is in a unique position to successfully implement and complete this research project due to its capabilities and experiences in funded research.

In 1985, the College of Pharmacy was awarded 5.3 million dollars (over 5 years) for a research center under NIH division of research resources.

A list of the major research and training grants in the College of Pharmacy and Pharmaceutical Sciences in Florida A&M University is available for your review upon request.

4. Length of Time Required to Complete the Research Project:

The project can be completed in a period of 2 years, after which the overall project will be evaluated and statistical analysis performed on the study group and the control group to determine program effectiveness. The students will be followed up for 2 additional years.

Literature Cited (under significance section)

1. National Cancer Institute. The Smoking Digest: Progress Report on a National Kicking the Habit. Bethesda, MD, October, 1977.

2. **Kuller, L., et al.,** Control of Cigarette Smoking From a Medical Perspective. In Annual Review of Public Health, edited by L. Breslow, J. Fielding and L. Lave. Vol. 3. Annual Reviews, Inc., Palo Alto, CA, 1982.

3. Vital Statistics of United States 1976: Mortality. U.S. Department of Health and Human Services, 1980, Vol. II, p. 194.

4. Vital Statistics of United States 1967: Mortality. U.S. Department of Health, Education and Welfare, 1970. Vol. II, p. 194.

5. **Williams, J. D. (ed.).** The State of Black American 1980. New York: National Urban League, 1980, p. 63—64.

6. **Leventhal, H., and Cleary, P.** The Smoking Problem: A Review of the Research and Theory in Behavioral Risk Modification. Psychol. Bull. 88: 370—405, 1981.

7. **Benfari, R. C., Ockene, J. K., and McIntyre, K. M.,** Control of Cigarette Smoking from a Psychological Perspective. In Annual Review of Public Health, Edited by L. Bes, J. Fielding and L. Lave.

8. **Russell, M. A. H., et al.,** Effect of General Practitioners' Advice Against Smoking. Brit. Med. J. 2: 231—235, 1979.

9. **Glasgow, R. E., et al.,** Self-help Books and Amount of Therapist Contact in Smoking Cessation Programs. J. Consult. Clin. Psychol. 48: 659—667, 1981.

10. **Shavfer, H., et al.,** The Primary Prevention of Smoking on Set: An Innoculation Approach. J. Psychoactive Drugs, 15(3): 177—184, July—September 1983.

11. **Choy, D. S., et al.,** Effective Treatment for Smoking Cessation. Am. J. Med. 75(6): 1033—1036, Dec. 1983.

12. **Stachnik, T. J., et al.,** Is There A Future for Smoking Cessation Programs? J. Community Health, 7(1): 47—56, Fall, 1981.

13. **Lotecka, et al.,** Enhancing Decision Behavior in High School Smokers. Int. J. Addit., 18(4): 479—490, May 1983.

14. **Schinke, S. P., et al.,** Primary Prevention of Tobacco Smoking, J. Sch. Health, 53(7): 416, Sept. 1983.

15. **Perry, C. L., et al.,** High School Smoking Prevention: The Relative Efficacy of Varied Treatment and Instructors. Adolescence, 18(71): 561—566, Fall 1983.

16. **Kendel, D. B., Logan, J. A.,** "Patterns of Drug Use from Adolescence to Young Adulthood": I. Periods of Risk for Initiation, Continued Use and Discontinuation. Am. J. of Public Health 74(7): 660—666, July 1984.

17. **O'Rourke, T., Smith, B. J., Nolte, A. E.,** "Health Risk Attitudes, Beliefs and Behaviors of Students Grades 7-12". J. of Sch. Health 54(5): 210—214, May 1984.

18. **Oei, T. P. S., Brasch, P., Silva, P. A.,** "The Prevalence of Smoking Among Dunedin Nine-Years-Old:. New Zealand Med. J. 97(761): 528—531, Aug. 8, 1984.

19. Lancet. "Young Smokers" Lancet 2(8469/8470): 1442—1443, Dec. 21-28, 1985.

20. **Murray, M., Swan, A. V., Clarke, G.,** "Long Term Effect of a School Based Antismoking Programme". J. Epidemiology and Community Health 38(3): 247—252, Sept. 1984.

21. **Tell, G. S., Klepp, K. I., Vellar, O. D. et al.,** "Preventing the Onset of Cigarette Smoking in Norwegian Adolescents: The Oslo Youth Study". Preventive Med. 13(3): 256—275, May 1984.

22. **Calman, A. L. H., Carmichael, S., Deans, H. G., et al.,** "Development of a Primary School Health Education Programme With Special Emphasis on the Prevention of Cigarette Smoking", Health Education J. 44: 65—69, 1985.

23. **Best, S. A., Flay, B. R., Towson, S. M. J., et al.,** "Smoking Prevention and the Concept of Risk" J. of Applied Soc. Psychology 14(3): 257—273, 1984.

24. **Christenuson, P. D.,** "An Evaluation of Quantitative and Qualitative Assessment Techniques on the Impact of the SERR Curriculum", J. of Sch. Health 55(5): 200—204, May 1985.

25. **Murray, D. M., Luepker, R. V., Johnson, C. A., et al.,** "The Prevention of Cigarette Smoking in Children: A Comparison of Four Strategies" J. of Applied Soc. Psy. 14(3): 274—288, 1984.

26. **Schinke, S. P., Gilchrist, L. D.,** "Preventing Cigarette Smoking with Youth". J. of Primary Prevention 5(1): 48—56, Fall 1984.

27. **Walter, H. J., Hofman, A., connelly, P. A., et al.,** "Primary Prevention of Chronic Disease in Childhood: Changes in Risk Factors After One Year of Intervention". Am. J. of Ep. 122(5): 772-781, Nov. 1985.

28. **Best, J. A., Flay, B. B.,** "The Waterloo Smoking Prevention Project: Methodological Advances, Results and Empirical Guidelines for Dissemination to the Schools", In: Forbes, W. F., Frecker, R. C., Nostbakken, D. (Editors). Proceedings of the Fifth World Conference on Smoking and Health, Winnipeg, Canada 1983, Vol. 1. Ottawa Canadian Council on Smoking and Health 1983, p. 649—655.

29. **Christenson, P. D.,** "An Evaluation of Quantitative and Qualitative Assessment Techniques on the Impact of the SERR Curriculum". J. of Sch. Health 55(5): 200—204, May 1985.

30. **Jones, A. W.,** "An Exploratory Study of the Etiology of Cigarette Smoking Among Kindergarten through Fourth-Grade School Children". Doctoral Dissertation, Pennsylvania State University, College of Health, Physical Education, and Recreation, University Microfilms International 85-15123, May 1985, p. 193.

31. **Epslein, L., Tamir, A.,** "Behavioral Risk Factors in Adolescence as a Basis for Health Planning" Israel J. of Med. Sci. 21(7): 595—600, July 1985.

32. **Murray, M., Kiryluk, S., Swan, A. V.,** "Relation Between Parents' and Childrens' Smoking Behavior and Attitudes". J. of Epi. and Community Health 39(2): 169—174, June 1985.

33. **Allendorff, S., Sunseri, A. J., Cullinan, J., et al.,** "Student Heart Health Knowledge Smoking Attitudes and Self-esteem". J. of Sch. Health 55(5): 196—199, May 1985.

34. **Croft, J. B., Hunter, S. M., Webber, L. S., et al.,** "Cigarette Smoking Behavior Distinctions Between Experimental Non-Smokers and Adopters in Children and Adolescents — A Consideration of Transitional Smoking Experience: The Bogalusa Heart Study". Preventive Med. 14(1): 109—122, Jan. 1985.

35. **Dignan, M. D., Steckler, A., Block, G. D., et al.,** "Prevalence of Health Risk Behavior Among Seventh Grade Students in North Carolina". Southern Med. J. 72(3): 295—302, March 1986.

1. Letter Regarding Decision on the Research Proposal

10 April 1989

College of Pharmacy & Pharmaceutical Sciences
Florida Agricultural & Mechanical University
Tallahassee, FL 32307

Dear   RE: Application PBR-

This is to inform you that the application to the American Cancer Society from Florida Agricultural & Mechanical University, received for the 1 November 1988 deadline, to support your project entitled, ''Smoking Prevention in Black Children'', has been approved by our Scientific Advisory committees, but was assigned a priority rating below the level we expect will be approved for funding by our Board of Directors at its meeting on 9 June 1989.

You will receive an official letter shortly after our Board of Directors' meeting indicating the final action taken on your application.

Sincerely,

Vice President for Research

2. Follow-Up Memo Regarding Request for Additional Information

TO: [redacted]

FROM: [redacted]
VICE PRESIDENT FOR RESEARCH

THIS IS IN RESPONSE TO YOUR RECENT REQUEST FOR ADDITIONAL INFORMATION REGARDING THE EVALUATION OF YOUR GRANT APPLICATION.

ATTACHED YOU WILL FIND A COPY OF THE COMMENTS MADE BY OUR SCIENTIFIC ADVISORY COMMITTEES. I HOPE THIS INFORMATION WILL BE OF SOME ASSISTANCE.

D. REVIEWER COMMENTS AND CRITIQUE

Objectives

The investigators propose to study interventions to deter smoking among black children in a university-related elementary school. In the background are the well-known risks of smoking, the higher risks among blacks than whites and the importance of early behavior on smoking prevalence. The investigators will use programmatic materials from the American Lung Association and other sources, including posters, films and puzzles in a one-hour-per-week intervention during the school years from grades 4 through 7. Pre and post tests in the intervention classes will be compared with tests in ''comparable'' black elementary schools in the community. The investigators' objective is to reduce smoking start-ups among the target children. Their aims are to test effectiveness of the interventions and to find whether parents and family are influenced.

Critique

On the plus side, the investigators are apparently in a position to reach the students in this university-affiliated school and have educational contacts and experience to conduct the research. The objectives are very important and the research is in a high-priority area.

On the negative side, reach is exceeding grasp. The project is very ambitious, proposes numerous questionnaires and time-consuming interventions, but the budget seems miniscule for the effort of a team needed to accomplish such goals. Specific problems include:

1. There are no specifics about numbers of children — and their teachers — who would be involved. Therefore, it is impossible to know whether any changes measured would be significant.
2. The interventions are mostly for older children; are they usable for 4th graders? Also, a detailed plan for intervention is not provided.
3. The stability of the study population is only briefly mentioned. In a 4 year study, drop-outs and turnovers might be expected to be very large.
4. There is no mention of validation procedures planned relative to smoking status of participants.

Chapter 2

THE NATURE AND PROCESS OF DOING RESEARCH

M. Montagne, H. F. Kabat, and E. Nnadi Okolo

TABLE OF CONTENTS

* M. Montagne.

** H. F. Kabat.

I. INTRODUCTION

Research begins with a desire or a need to study a phenomenon, problem, intervention, or event. Research, by its very nature, is a continuous process that is a part of social and technological contexts. The process of doing research involves many interdependent activities, always in connection with the "real" world. The purpose of doing most research is to generate knowledge, which should be viewed as a social commodity to be shared and used.[1] The aim of scientific research is to make discoveries, about us and our world, although to some degree every discovery disturbs accepted ideas and the current state of knowledge.[2] The key elements of science, and of doing research, are curiosity, veracity, and the hard work of testing ideas in an empirical manner.

II. THE RESEARCH ACT

Scientific research is one way of generating and testing statements about events and human experiences. At least four ways of generating and testing the "truth" of empirical statements have been identified: the authoritarian, the mystical, the logico-rational, and the scientific.[3] One approach to generating knowledge does not necessarily preclude the usc of the others in the same inquiry. In the authoritarian approach, knowledge is sought and tested by referring to those who are socially defined as qualified producers of knowledge. The mystical approach is a variation of the authoritarian one. While the mystical approach also depends on individuals or entities in natural or supernatural positions of authority, the generation and testing of knowledge is based more on ritualistic procedures and may rely on more abstract, magical, or inspirational sources of information.

In the logico-rational approach, knowledge is generated through a procedure that employs the rules of formal logic. The scientific approach is reliant primarily on the observation of the phenomenon or experience to generate knowledge about it, and it is less reliant on the actual procedures used to generate that knowledge. It should be realized that "all of the methods of science may be thought of as relatively strict cultural conventions whereby the production, transformation, and therefore the criticism, of proposed items of knowledge may be carried out collectively and with relatively unequivocal results."[3] Scientific methods are, quite simply, the ways in which agreement about phenomena and experiences in our world are reached.

A. HISTORY AND PHILOSOPHY OF SCIENTIFIC ENDEAVORS

There is a basic assumption that the empirical world exists as something to be observed and studied, which is the raison d'etre for the existence of a scientific discipline. It should be noted, however, that the philosophical positions of realism and idealism can profoundly influence the nature of scientific inquiry. The philosophical view of some researchers states that the empirical world exists only in human experience, and it appears only in a form that humans can observe and describe. Many scientists would argue, however, that this viewpoint is solipsistic and untenable. Realists, on the other hand, argue that the objects of human cognition actually do exist; they claim to depict, through their research, the facts or phenomena as they really are. However, realism is often viewed as being fixed and immutable, which can limit its usefulness in describing and explaining phenomena. This belief can be perpetuated by activity in a specific scientific field, such as the physical sciences, which can extend to other scientific endeavors. It should be realized that "the history of empirical science shows that the reality of the empirical world appears in the 'here and now' and is continuously recast with the achievement of new discoveries."[4] The goal of science is to devise images of the world and to test them; methods are the principles that guide this process.

The standard view of the development of scientific knowledge is logical positivism or, as it is sometimes called, rationalism. This dominant paradigm of science abandons the search for ultimate causes and limits knowledge to the factual character of real-world observations. This type of knowledge, from this perspective, is the only type that is both attainable and useful. There is a tendency in the positivistic approach to reduce scientific knowledge and methods to the purely objective observation and classification of phenomena as they are actually experienced. In this approach, the research endeavor starts with a problem, which the researcher then attempts to solve in a deductive manner.[5,6] The route to a solution involves developing a theory with testable hypotheses generated *a priori*. There is a methodological insistence on controlled and independent observations of real events or phenomena as the basis for accepting or rejecting claims about the truth of a particular bit of knowledge.

There is also a viewpoint that science is an essentially anarchistic enterprise and that it might be closer to myth than many are willing to admit.[7] Methodologies are seen as having their limits; in fact, the only principle that apparently does not inhibit progress in scientific endeavors is the notion that *anything goes*. This anarchistic epistemology suggests that the idea of doing research according to fixed and universal rules is unrealistic. This perspective is based in part on larger political and social concerns regarding the place of science in society and the practical application of scientific knowledge to everyday life.

B. THE EPISTEMOLOGY OF DOING RESEARCH

The publication of *The Structure of Scientific Revolutions* challenged the commonly held assumption that science advances in a cumulative manner, with each advance in knowledge building on the knowledge and ideas that had preceded it. In other words, scientific knowledge was viewed, and still is by many, as being built slowly and steadily in an incremental manner. Instead, as Kuhn[8,9] argued, while accumulation plays a role in the advancement of scientific thought, the truly major changes come about in the form of revolutions. In his view, a scientific discipline at any given point in time is dominated by a specific paradigm. A paradigm can be defined as the fundamental image, or world view, of that science's subject matter — what it is and how it should be studied. There is a period of time over which that discipline accumulates knowledge, resulting in an expansion and clarification of its theories and practices. Such work inevitably generates anomalies or information and concepts that cannot be explained by the existing paradigm. These anomalies can build up to a point where a crisis within that discipline may occur and a revolution in ideas takes place. A new paradigm will result. In a political sense, a paradigm is the major dimension for defining one scientific discipline in contradistinction to another.

A scientific discipline is at its peak when its paradigm has gained hegemony within a broader scientific field. The discipline seeks to solidify and to expand its position by (1) extending the knowledge of those facts that the paradigm displays as particularly revealing, (2) increasing the extent of the match between those facts and the paradigm's predictions, and (3) furthering the articulation of the paradigm itself.[8] While at its peak the discipline is in control of extending factual knowledge, but most often major novelties of a conceptual or phenomenal nature are not generated. In the course of extending this knowledge, research will uncover facts or observations that are not congruent with the dominant paradigm. In order to accommodate these anomalies into existing knowledge, a new paradigm will likely emerge. This new paradigm must be different from the previous one, which it is replacing, in the sense that it seems to explain the phenomenon more clearly and demonstrably. It is during this shift in paradigms that the greatest changes in science occur. New concepts and theories are defined, different questions are asked, and innovative methodologies are developed. This idea of the development of a science is very different from the notion that science develops, for the most part, in a cumulative manner.

C. MAJOR RESEARCH PARADIGMS

There are many approaches, or paradigms, to doing research. These different varieties are comparable along many continua (e.g., qualitative-quantitative, deductive-inductive, and so forth). The primary paradigmatic comparison seems to be with or against the currently dominant approach, that being the classic hypothetico-deductive paradigm, which was developed in the natural sciences. The exemplar of the hypothetico-deductive approach stresses experimental or quasi-experimental designs with hypotheses being generated *a priori* and tested quantitatively.[10-13] The data are collected with structured, or even standardized, instruments developed from specifically stated research goals or hypotheses. These data are usually analyzed with multivariate, parametric statistical techniques. In essence, the researcher is interested in collecting numerical types of information. Results are expressed in terms of statistically significant effects according to the stated hypotheses. The hypothetico-deductive paradigm has been called "the only available route to cumulative progress".[11]

The major alternate approach, developed in the social sciences, has been referred to as holistic-inductive in nature.[4,14,15] This approach employs qualitative methods such as field observations and small group analysis in deriving a detailed description of the characteristics of an event, phenomenon, or experience. This approach trades off the testing of *a priori* hypotheses and some measurement precision in order to obtain richer and more detailed information. The techniques of qualitative methodology have been described as being just as objective, useful, specific, and important as quantitative methods.[16,17]

Either of the two aforementioned approaches is a general perspective, or a paradigm, on viewing and describing the complexities of the real world. These approaches are deeply embedded in the socialization of adherents and researchers; it tells them what is important, legitimate, and even what should be studied. These paradigms are normative; they tell us how to do research without the necessity of long existential or epistemological considerations. However, it is this aspect of paradigms that constitutes both their strength and weakness — their strength in that they make action possible; their weakness in that the very reason for action is hidden in the unquestioned assumptions of that paradigm. The point here is that different kinds of research problems require different types of epistemologies and methodologies to generate useful information. The hypothetico-deductive paradigm, employing quantitative methods, aims at describing the causes or consequences of a phenomenon or experience, while the holistic-inductive paradigm, employing qualitative methods, aims at describing the characteristics of a phenomenon or experience. Each approach has its useful and important place in research endeavors.

D. ETHICAL CONSIDERATIONS

There are three ethical principles that form the basis for health care practice and research. These principles are a major part of the foundation for ethics in Western societies, and they are represented in Judeo-Christian religious beliefs and the Hippocratic tradition.[18,19] The rule of "Do No Harm" reflects the principle of beneficence. Adherence to this principle requires that good be done, that harm be avoided or minimized, and that the benefits of research justify any harm that may result. This principle dominates health care, as shown in the practice of prolonging life often through great physical and mental harm. Autonomy, or respect for a person, is a principle that focuses on the free choice of individuals. It requires that this right of individual decision-making be honored and that protection be given to those of diminished autonomy (e.g., youth, the mentally impaired, incarcerated individuals). In health care practice and research, this means that an individual has the right to consent or to refuse treatment. The last principle is that of justice or equity. It requires an equal distribution of benefits and burdens, except where an unequal distribution may be justified by merit or need.

Some ethicists include additional principles, and most of these are pertinent to health care research.[20,21] These principles often form the ethical basis of professional behaviors, and they are viewed as duties, including (1) fidelity in contract-keeping, or faithfulness to a covenant between those involved in a given action or decision; (2) veracity, or telling the truth and not deceiving in any way; (3) the avoidance of killing; (4) a respect for nature; (5) the provision and maintenance of privacy and confidentiality; and (6) a consideration of potential conflicts in contractual and social role obligations. In a particular situation or ethical dilemma there are a number of approaches that can be taken. One approach is to rank order these principles in some manner and to give priority to some over others. In this approach, an individual might follow a single principle, such as beneficence or "Do No Harm", and allow it to override others in all actions and decisions. A very different approach would attempt to balance the various principles, giving relatively equal consideration to all of them in a given instance. This latter approach, often called situational ethics, is the more popular approach in our society. In addition, laws, regulations, and social norms often direct or dictate which ethical principle should predominate.

Another way of standardizing and employing ethical principles involves codes of ethics. The most familiar code for most of medical history is the oath adopted by the school of the physician Hippocrates in the 5th century B.C. It has been, and continues to be, the primary code for directing appropriate moral actions in health care. With the emergence of scientific medicine in the 19th century, new codes of ethics were formulated. The first general codification of medical ethics in the U.S. was developed by the American Medical Association in 1848, and that for the pharmaceutical profession was developed by the Philadelphia College of Pharmacy in that same year.[22] The key principles in these codes were a respect for human life and service to mankind.

The major development in ethical codes for health care research came after the Second World War. Biomedical experimentation on humans during wartime and judgments handed down at the Nuremberg trials led to the creation of the Code of Nuremberg.[23] This code was designed to regulate biomedical research. It is based on three principles: (1) obligations to the individual subjects involved in the research experiment must be placed above obligations to the state (governments), (2) the distinction between therapeutic and nontherapeutic research is taken to have moral implications, and (3) informed consent from the subject is morally essential. This code dictated that consent to participate in medical treatment or research must be completely voluntary, and it must be obtained in terms of a full and enlightened understanding of all aspects of the treatment or research experiment, most especially its hazards. In the particular case of research, this code stated that experiments should seek a benefit for society that is unprocurable by any other means, that it should be based on prior studies in animals, and that all unnecessary physical and mental risks should be minimized.[24]

Around this same time the World Medical Association updated the Hippocratic Oath, and in 1949 it established the Geneva Convention Code of Medical Ethics.[23] This code directed physicians to consecrate their lives to the service of humanity and to maintain the utmost respect for human life from the time of conception, even under threat. In the 1960s the World Medical Association adopted a code of ethics for experimentation with human subjects. The Declaration of Helsinki (1964, revised 1975) incorporated the three basic ethical principles from the Nuremberg Code, and it made a fundamental distinction between research that is essentially therapeutic for the patient and research that is purely scientific, with little or no direct therapeutic benefit to the patient. Ethical principles in the provision of medical care were viewed as the foundation for ethical principles which guide health care research.[25] The Declaration of Hawaii (1977) was even more specific in focusing on psychiatry and psychiatric research.[23] Its articles forbid the use of the resources of psychiatry for nontherapeutic reasons, most especially political or ideological ones.

As a result of these developments, a research ethic has evolved which is the "deliberate and systematic consideration of moral problems arising in connection with the conduct and consequence of scientific research''.[23] It is relatively easy to develop and establish ethical codes in health care practice and research, but it is often more difficult to use or apply these codes and principles in real life situations. Ethical principles in health care practice and research should be viewed in the context of an overall order of ethical priorities in life. The use of ethical principles in providing medical treatment should not be viewed as superior to or independent of the general ethics of living, and ethics in health care research are not superior to or independent of medical ethics. The formulation of ethical codes and oaths does not negate the need to carefully consider alternative actions and decisions that can be made in individual situations. Health care research can be conducted in an ethical manner if the experiment is carefully described to the participants in a way that clearly distinguishes the therapeutic components from the research ones. The cornerstone to the ethical design and conduct of health care research is informed consent.

E. PATIENT SELECTION AND INFORMED CONSENT

In the early 1980s the general public's attitude toward clinical trials was positive.[26] Participation in clinical trials was seen as benefiting patients and society by helping to improve medical knowledge. The public was still concerned, however, about the use of placebo controls in experiments, determinations of when it was permissible to withhold standard treatment, and the lengthy time period it took to apply experimental results in general medical practice. For the most part the general public was supportive of clinical trials, and most assumed that medical research in the U.S. was always ethical. The public's strong support was due, in part, to the development over the 1960s and 1970s of informed consent procedures.[27] Informed consent evolved from the Nuremberg Code and one of its key principles, that voluntary consent of the human subject is morally essential.

However, consent procedures have had an influence on the process of doing research.[28,29] For instance, the randomization procedure in research studies is the focal point of many concerns, and there are even different viewpoints from the researcher's perspective.[30] One view holds that full disclosure is necessary since every subject is legally and morally entitled to undertake his or her own evaluation of the risks and benefits of participation in the trial, and full knowledge is needed in order to make such a decision. A different view holds that disclosure of some information, particularly the randomization process, is not necessary since there is nothing material to disclose so long as the ''better'' treatment is really unknown and will not be known until the trial is completed and the data are analyzed. These concerns have been further highlighted by the acquired immune deficiency syndrome (AIDS) epidemic of the 1980s and the search for effective substances to treat that disease.

The current consensus is, however, that the most optimal and ethical method for investigation of health care problems and issues is the randomized clinical trial.[31] An experiment is inherently ethical or unethical in its design and execution, and most promising ends cannot justify unethical means. The researcher or clinician can be presented, in a clinical trial, with a potential conflict of values between his/her role as healer and his/her role as scientist or between the goals of therapy and the goals of experimentation. The physician may not be able to commit exclusively and unequivocally to the interests of the patient. To counter some of these concerns and potential problems, the use of institutional review boards, to approve study designs and assure for informed consent, and advisory committees, for monitoring ongoing study results and informing investigators whether the new treatment is significantly superior or inferior to the alternative treatments, have been advocated.

In response to concerns about the ethics of randomization procedures in research, a new design for planning randomized clinical trials has been proposed.[32] This method is well suited to comparisons of standard or control treatment with an experimental treatment. In this design, patients are randomized into two groups. Patients in the first group receive the standard treatment. Patients in the second group are asked if they will accept the experimental treatment; if they decline, they receive the standard therapy. In analyzing the data at the end of the trial, all patients in the second group, regardless of treatment, are compared with those in the first group. Any loss of statistical efficiency may be overcome by increasing the number of patients. The presumed advantage of this design is that before providing consent the patient will know if an experimental procedure is to be used or not. This proposed design, however, has received much commentary and criticism. It is acknowledged as ethical and reasonable since it removes the patient from unnecessary participation in the randomization process and allows some patients to choose between standard and experimental treatments. The design has disadvantages, though, in eliminating blinding and reducing statistical efficiency, and it does raise different ethical issues.

Informed consent implies the disclosure and comprehension of information about the research study or experiment and the voluntary consent of the subject while he or she is competent.[33] The concept of informed consent addresses the three basic ethical principles, most especially the notion of autonomy. A patient's or subject's signature on a consent form or vocal approval of consent is, however, no assurance that the person has truly given informed consent. Some subjects have been known to use the research process for their own benefits, while other subjects are used by it.[28] The development and implementation of institutional review boards and committees on experimentation in human subjects is one method for assuring integrity in informed consent and ethical practices in the whole process of doing research.[34] The ability of these boards to function and to monitor research and their effectiveness in assuring informed consent may actually be limited in real life settings.[35]

1. Basic Elements

There are a number of basic elements to informed consent agreements with human subjects.[29] The essential aspects of the research process that must be presented to potential participants include:

1. An invitation, not a request or a demand, to participate
2. A description of the research project:
   a. Title of the project
   b. Purpose of the research project
   c. The experimental/study procedures
   d. Location where the study will be conducted
   e. The basis for selecting certain patients
   f. Duration and scheduling of patient participation
3. A description of foreseeable risks and discomforts to the patient
4. A listing of appropriate alternative treatments and procedures that may benefit and that may be available to the patient
5. A description of the benefits of the study directly to the individual patient, and to other patients and society
6. A statement regarding confidentiality of patient records, or in some instances (such as in some survey research) a description of how the participant's anonymity will be assured and maintained
7. An explanation of compensation or medical treatment provided to the patient in case of injury or harm
8. A description of procedures used in storing and destroying all materials and samples at the conclusion of the study
9. A listing of the names and phone numbers of the principal investigator, contact persons, and/or other personnel connected with the study, in case the participant has questions, concerns, problems, or wishes to withdraw
10. A statement making it clear to the potential participant that participation is voluntary, that a choice not to participate will involve no penalty or change in current or future treatment, and that the patient may withdraw at any time without penalty or loss

Other pertinent information is oftentimes included on the informed consent form, or it is offered to the participants, including (1) a statement of risks to an embryo or fetus; (2) a description of termination of treatment by the principal investigator without regard for patient consent; (3) a listing of additional costs due to participation; (4) an outline of procedures and consequences should a patient decide to withdraw, so that an orderly termination can occur; (5) a statement of the total number of participants expected to be involved in the study; and (6) agreements to provide new information, which has emerged from the study itself, to the patients during the course of study. In most cases, the inclusion of these items and points of information are greatly dependent upon the guidelines of individual institutions and their review boards, or they result from changes in governmental policies regarding the use of human subjects in medical research. Health care researchers should always refer to the specific policies and procedures of their affiliated institutions, and from time to time they should also check for changes in policies at the level of professional organizations and governmental agencies.

In addition to these basic ethical principles, which apply to most persons in the general population, there are special concerns that are related to research involving the embryo and fetus, children, the mentally infirm, and prisoners.[29] Most of these concerns revolve around issues of autonomy and, to some extent, justice or equity, especially when these individuals are stereotyped by investigators as being passive or inferior in some way. There are also practical and technical issues and problems (e.g., defining risks, obtaining informed consent) that arise in research involving these special population subgroups. Many of these issues or problems are reasons, or rationales, for not performing research on these subgroups. This can present an additional problem of limiting scientific and clinical knowledge about these groups of people, which may limit the quality and currency of the health care that they receive.

III. THE PROCESS OF DOING RESEARCH

Scientific methods are the ways in which researchers organize their inquiries so that they can be explicit and systematic in reporting their experiences. It is the way of moving from observation to conclusions. Within the scientific community there are a great variety of methodological approaches. There are some methods that work well in the research activities of one discipline, but are not useful for another field of study. Observation, experimentation, and other research strategies for collecting information can be viewed as specific methods.[36] These should not be considered as incompatible, for some disciplines can use many different methodological approaches, alone and in combination. Sometimes a particular discipline assumes a particular method as representative of its epistemology. Measurement and logic can be viewed as the general methods of science; they are, to varying degrees, a part of the process of doing research in any and all scientific disciplines. However, it has been noted that ''many scientific methods are not peculiar to science. The alchemists had experiment, the astrologers had careful observation, the geomancers and diviners had measurement, the theologians had logic.''[36]

The research process can be divided into the acts of theorizing, or constructing theory, and of empirical inquiry. Two methodological stances are also involved. The inductive approach provides an understanding of what is observed and assists in constructing theory, while the deductive approach provides guidance in knowing what to observe and assists in testing and applying theory.

A. FORMULATING A PROBLEM FOR SCIENTIFIC INQUIRY

The operationist school of thought believes that no concept or scientific term can be significant unless it possesses an empirical interpretation. In a general sense, the concepts or terms used to describe an experience or phenomenon must be framed in terms of activities, or ''operations'', that can be performed clearly. This would include the process of defining and specifying the component parts and relations of the concept, as well as the methods to measure it. Instead of relying solely on mental operations to direct data collection and analysis, there should be a ''consideration that scientific terms are to function in statements which are capable of objective test by reference to data furnished by direct observation. 'Operational definition' was conceived as a means to insure their suitability for this purpose by providing criteria for the test of the statements in which the terms occur.''[37]

This empirical interpretation can turn a conceptual system into a testable theory. The utility and testability of a theory, however, are not based solely on fully operational definitions for all constituent concepts. Each statement of empirical science should be formulated with ''the recognition that a scientific hypothesis cannot, as a rule, be tested in isolation but only in combination with other statements, and by replacing the overly rigid standard of complete verifiability or falsifiability by the more liberal requirement that a system of hypotheses must be capable of being more or less highly confirmed by observational evidence.''[37]

B. THEORY CONSTRUCTION

In attempting to describe or interpret the event, phenomenon, or experience under study, a researcher will oftentimes construct a theory. A number of approaches to theory construction have been developed.[38-40] There are at least two basic strategies for constructing a theory, as described by Freese.[41] A generalizing strategy has the following features: ''Its objectives are to explain and to generalize about the lawful phenomenon of open systems; its structure is systematic and it contains ordinary language; its method is inductive abstraction; and its possible result is the consolidation of theories or data.''[41] On the other hand, a pure, or instrumental, strategy has the following features: ''Its objective is to predict the behavior of lawful phenomena in closed sytems; its structure is formal and it contains no ordinary language; its method is idealization; and its possible result is the cumulation of theory.''[41] These two strategies are not mutually exclusive, and many have argued that in practice a mixture of both types occurs.

A theory, then, represents a series of interrelated concepts. It is usually formulated in terms of statements or propositions that can be verified though empirical testing. It may be applied to explain or predict an event, phenomenon, or experience. The testing of a theory is done by identifying and defining the theoretical concept, specifying the variables which comprise that concept, and converting these variables into an operational form for measurement. There are some interesting points about theory construction that present it more as science and less as magic, at least from a social perspective. First, not all ideas about natural and social phenomena are theories. Theories, in the context of this brief overview, are those ideas that are formally defined and constructed for the purpose of scientific investigation. Theories are also not composed by ideas about history; rather, they represent an attempt to truly reflect ''reality'', and as a result they can always be improved. Theories rarely, and some would say never, deal with axiological judgments (i.e., what should be); rather, they deal with existential judgments (i.e., what is). Of course, theories should not exist as untested ideas, but should be tested in some form.

Theories take a certain form and a number of different types have been described, but first there are differences in the basic relationship between theorizing and doing research.[39] The more popular and traditional (as currently viewed) approach has been described as ''theory then research''. The major steps of this strategy are (1) the development of an explicit theory in an abstract structured form, (2) the testing of statements generated by the theory through empirical research, and (3) the refinement of the theory and further testing of statements to determine the strengths and limitations of the theory. The alternate strategy offers the process of (1) selecting a phenomenon or experience for study and determining its characteristics, (2) measuring these characteristics in a number of different situations, and (3) analyzing this information and formalizing any significant patterns or trends in terms of theoretical statements. This approach of generating theory, after data collection, from information that has been obtained systematically in a research endeavor is often referred to as the discovery of grounded theory.[14]

1. Theory-then-Research

Within the theory-then-research strategy there are three ways of organizing sets of statements into theoretical form.[39] The *set-of-laws form* allows for the construction of scientific knowledge through a set of theoretical statements that have overwhelming empirical support. The *axiomatic form* is defined as an interlinked set of concepts, definitions, and statements. In this form, the theoretical concepts are identified and defined operationally, statements are constructed to describe situations in which the theory can be applied, and another set of statements is generated to describe the interrelationships between concepts and axiomatic statements. A system of logic is also used to relate all concepts within the statements and to derive propositions for scientific testing. The final form constructed theories might take describes *causal processes*. In this format, a set of statements is organized in such a way that the causal relationships between two or more concepts are made explicit.

The two key components of most theories are concepts and statements. Concepts may be specific and concrete, or they may be abstract in nature. Abstract concepts are completely independent of a specific time, place, or individual. As these factors are specified and defined for a certain situation, the concepts become less abstract. A description of a relationship between concepts is called a statement, and its level of abstractness will depend upon the abstract or concrete nature of the concepts of which it is comprised. These concepts then become operationalized, wherein definitions are generated to describe the testing or empirical determination of a theoretical concept in specified circumstances.

Statements comprise the "structure" of a theory. Theoretical statements have been classified into two types.[39] Existence statements make claims about a phenomenon, as described by the theoretical concepts, while relational statements describe the linkages between those concepts. Theoretical statements about the existence of a phenomenon or experience can be framed in terms of laws, axioms, propositions, hypotheses, or empirical generalizations. Theoretical statements about the interrelationships between concepts can describe correlation or causality. Finally, assumptions are things taken for granted about the empirical world. They are often statements that are taken as given. They are rarely subjected to direct empirical verification or testing. This is, however, certainly an arguable point since assumptions are usually made when something is not amenable to empirical testing or when the present state of knowledge does not allow for proving or disproving it.

The most interesting and useful statements, in terms of research applications, are hypotheses.[40] Hypotheses are generated from the theory. They are essentially statements, most often of a relational type, of what ought to be observed if the theory itself is "correct" or empirically tenable. Hypotheses are also viewed as pretheoretical propositions which indicate a relationship between at least two facts, aspects, items, or phenomena. Hypotheses are formally structured in one of two ways. The traditionally "ideal" hypothesis takes the form of an if-then statement. The "if" part specifies the conditions which must exist if the hypothesis is to be applicable. It subsumes the factors and contexts from which a prediction is being made and grounded. The "then" part predicts what is expected to occur when the conditions exist. Hypotheses also may be stated in the form of a continuous connection. These types of hypotheses are often stated in mathematical terms, and they imply degrees of correlation as well as causality.

C. BEGINNING THE RESEARCH PROCESS

The conceptualization of a research plan can be done in a rather simple and straightforward manner, based upon set rules and guidelines developed within a particular scientific discipline. Much of this book on health research and design will provide and describe specific examples of such plans. It should be realized that paradigms for doing research exist, and these can be very influential in directing thought and action in the research process. An awareness of these different views of the research process is important so that the most appropriate perspective and method are chosen in studying a given phenomenon or experience. Perhaps the best overall approach is to "respect the nature of the empirical world and organize a methodological stance to reflect that respect".[4] It is the scientist's responsibility to adhere to the general principles of doing research and to always keep in mind the variety of social, political, and ideological factors that can affect the research process. In accepting his Nobel Prize in physics in 1946, Percy Bridgman stated that "there is no scientific method as such, but the vital feature of the scientist's procedure has been merely to do his upmost with his mind, *no holds barred*."[42]

1. Defining the Problem

Health research investigators seek to establish, through testing, the relationship between two or more variables, concepts, phenomena, events, or things that are measurable. The initial step in defining the problem to be investigated is to develop a one-sentence statement in explicit terms. "The purpose of this investigation is to establish the relationship between X and Y." This facilitates the development and clarification of a complementary one-sentence statement of the primary research question, "If there is a change in X, will there also be a concomitant change in Y?", which also may be stated as a hypothesis: "An increase in X will be accompanied by a concomitant increase in Y." There is a very close affinity between the statement of the problem, the primary research question, and the resulting hypothesis.

This exercise in clarification allows the investigator to pursue a variety of possible research directions while forcing thought, perspective, and imagination to the subject under investigation. Although the development of the one-sentence statement of problem and the primary research question may appear to be a fairly simple, straightforward process, it is rather the product of vigorous intellectual effort which, when successfully accomplished, results in a general falling into place of the broader research problem, the subsidiary questions, and even the required analytical technique. Every effort should be made to keep them simple by avoiding confusing, confounding, or multiple-subject statements, rather focusing them specifically to the health research subject under investigation.

Each important word or term must be defined to make the statement of the problem and the primary research question operational for purposes of the investigation. Any term in the one-sentence statement of the problem and the primary research question that may be subject to question or misinterpretation must be defined, as should all technical and semitechnical terms.

Once the statement of the problem and the research question have been explicitly stated and each major word or term adequately defined, related or subsidiary questions may be developed. They are identified separately from the primary research question. The purpose of the subsidiary question or hypothesis is to cast light on the primary research question. Each subsidiary question included should contribute something vital to resolution of the primary research question. The subsidiary questions help to refine the complexity of the relationships under investigation and may even help to determine the data that must be collected and the methods of analysis that will be used.

2. Literature Review

Once the research problem, primary research question or hypothesis, subsidiary questions or subhypotheses, and definitions of terms are in place, one can undertake a review of the research and theory literature relevant to the subject under investigation. When an investigator hypothesizes that "a change in X will be accompanied by a change in Y", there must be evidence or theory to directly support the hypothesis in the published literature. In addition, an investigator will wish to review all important literature that bears on the problem, primary research question, and subsidiary questions.

Health science research takes some investigators into the field, some into the laboratory, and almost all into the library at one time or another. Health science libraries offer a wide array of services and resources to assist with the review of the literature, not the least of which is the medical librarian, who has an intimate knowledge of the reference collection. The reference collection consists of bibliographic sources (monographs, periodicals, indexing and abstracting services, computerized data bases, U.S. Government documents, technical reports, conferences, reviews, and translations) and information sources (terminology, handbooks and manuals, drug information sources, audiovisual sources, medical and health statistics, directories, biographical sources, and history sources).

a. Monographs

Although monographs are no longer the primary means of printed communication in the health sciences, they are valuable sources for verification of bibliographic elements. Monographs in the reference collection are filed in the card catalog or computerized card catalog under author, title, subject, major topic headings, or divisions within a work. Current coverage of medical monographs is contained in the National Library of Medicine's (NLM) CATLINE data base, the NLM *Current Catalog and Current Catalog Proof Sheets* publications. The CATLINE data base is an on-line system from NLM that contains more than 600,000 citations dating back to 1965. It is the most up-to-date compilation of cataloging of materials in the health sciences. Medical monographs are also included in the *National Union Catalog*, *Cumulative Book Index*, *Weekly Record*, and the *American Book Publishing Record*.

b. Periodicals

The periodicals collection constitutes the major element in most health science libraries' holdings. The primary sources for current medical periodicals are the NLM SERLINE and CATLINE on-line systems and their published by-products, *Index of NLM Serial Titles*, *Health Science Serials*, *NLM Current Catalog*, and *List of Serials and Monographs Indexed for On-Line Users*. There are approximately 70,000 titles in the SERLINE data base. *Index Medicus*, *Excerpta Medica*, *BIOSIS*, *Science Citation Index*, and *Chemical Abstracts* list periodicals that they abstract.

c. Indexing and Abstracting Services

An index provides basic bibliographic information needed to locate an article, while the abstracting service also provides a brief summary of the article's content. Access is made by use of keywords, thesaurus terms, authors, or cited authors. The major index to medical periodical literature is *Index Medicus*. Other valuable indexes and abstracts are *Excerpta Medica*, *Medical Socioeconomic Research Sources*, *International Nursing Index*, *Cumulative Index to Nursing and Allied Health Literature*, *Nursing Studies Index*, *Index to Dental Literature*, *Dental Abstracts*, *Oral Research Abstracts*, *Hospital Literature Index*, *Hospital Abstracts*, *Abstracts of Health Care Management Studies*, and *International Pharmaceutical Abstracts*. Because there is a significant lag time between publication of a journal article and its appearance in an index or abstract, reference should be made to current awareness periodicals such as *Current Contents*, which appears in five editions: *Agriculture*, *Biology and Environmental Sciences*, *Social and Behavioral Sciences*, *Clinical Practice*, and *Life Sciences*.

d. Computerized Data Bases

Today, virtually every index and abstract service has been computerized, which permits on-line retrieval of bibliographic materials. In addition to access, computer technology makes it possible to manipulate the bibliographic material to extract citations for topics. The principal computer data base vendors are System Development Corporation's Search Service ORBIT, Lockheed's DIALOG, Bibliographic Retrieval Service (BRS), and the NLM. Each vendor provides detailed user's manuals which list the data bases and directions to search them. NLM data bases include MEDLINE, AIDSLINE, TOXLINE, CHEMLINE, RTECS, TDB, CANCERLIT, CANCERPRO, CCRIS, CLINPROT, HSDB, POPLINE, HEALTH PLANNING & ADMINISTRATION, BIOETHICSLINE, EPILEPSYLINE, Physician Data Query (PDQ), and HISTLINE. Non-NLM data bases include BIOSIS Previews, CA search, EMBASE, PSYCHOLOGICAL ABSTRACTS, SCISEARCH, Social SCISEARCH, Sociological Abstracts, SPORT, Pollution Abstracts, Drug Information Full Text (DIF), and IPA.

e. U.S. Government Documents and Technical Reports

Government documents appear in both print and nonprint formats. The primary bibliographies of U.S. government publications are the *Monthly Catalog of United States Government Publications* and MEDOC, a computerized index to U.S. Government documents in the medical and health sciences, which is on-line through BRS. The National Technical Information Service (NTIS) is the central source for public sale of government-sponsored research and development reports. Titles are listed in *Government Reports, Announcements & Index*, and *Weekly Government Abstracts*. The National Aeronautics and Space Administration publishes *Scientific and Technical Aerospace Reports*; the Energy Research and Development Administration publishes *Energy Research Abstract*; the Educational Resources Information Center publishes *Resources in Education*; and the British Library Lending Division publishes the *BLLD Announcement Bulletin*, listing available technical reports.

f. Conferences and Reviews

Meetings have become a major means for the exchange of scientific information with colleagues and for the establishment of lines of professional communication. *World Meetings: United States and Canada*, and *World Meetings: Outside United States and Canada* are the most comprehensive and detailed listings of future meetings in the sciences. Many professional organizations regularly publish the dates of future meetings in their disciplines. *Conference Paper Index* publishes a listing of all papers presented at meetings they have included. Among the current bibliographies of published proceedings of meetings are *Index to Scientific and Technical Proceedings*, *Directory of Published Proceedings*, *Series SEMT — Science/Engineering/Medicine/Technology*, and *Series PCE — Pollution Control Ecology*, *Proceedings in Print*, and *Index of Conference Proceedings Received*. *Bibliography of Medical Reviews* and *Index to Scientific Reviews* identify and publish reviews which synthesize information from a variety of sources.

g. Terminology

It is necessary to have a clear understanding of terminology and the tools that present the terminology in order to do research in health sciences. Among the most used medical dictionaries are Blakiston's, Gould, Dorland's Illustrated, and Steadman's. Abbreviations are detailed in *Medical Abbreviations and Acronyms*, *Dictionary of Abbreviations in Medicine and Related Sciences*, and *Dictionary of Abbreviations in Medicine and the Health Sciences*. *Origin of Medical Terms* and *A Sourcebook of Biological Names and Terms* are useful for medical etymology. Concept dictionaries and word finders include *Inverted Medical Dictionary*, *Medical Word Finder*, and *Reversicon: A Medical Word Finder*.

h. Handbooks and Manuals

These publications are compendia of vast amounts of information on a variety of subjects and are useful in addressing factual questions. Basic scientific information is presented in a concise format. Among the more useful are the *Biology Data Book*; *Growth, Human Health and Disease*; Ciba-Geigy's *Scientific Tables*; *Handbook of Chemistry and Physics*; *Composite Index for CRC Handbooks*; *Handbook of Biochemistry and Molecular Biology*; *Merck Index*; *Merck Manual of Diagnosis and Therapeutics: International Classification of Diseases*; and *Diagnostic and Statistical Manual of Mental Disorders*.

i. Drug Information Sources

Useful publications that address the action of drugs and chemicals on living systems include *Guide to Drug Information*, *American Drug Index*, *Merck Index*, *Unlisted Drugs*, *Martindale: The Extra Pharmacopoeia*, *The Pharmacological Basis of Therapeutics*, *Remington's Pharmaceutical Sciences*, *Physician's Desk Reference*, *PharmIndex*, *ToxIndex*, *AHFS' Drug Information*, *United States Pharmacopeia*, *Dispensing Information*, *AMA Drug Evaluations*, *Handbook of Nonprescription Drugs*, *Accepted Dental Therapeutics*, *Side Effects of Drugs*, *Evaluation of Drug Interactions*, Hansten's *Drug Interactions*, *Inpharma*, and *International Pharmaceutical Abstracts* (IPA).

j. Audiovisual Reference Sources

AVLINE, an on-line computerized data base maintained by NLM, provides access to more than 10,000 health science audiovisuals in all discipline areas. They also publish a catalog.

k. Medical and Health Statistics

Statistics are used to shed light on various aspects of the population, to understand how health services and health programs are being utilized, to evaluate the effectiveness of health care delivery, and to identify the country's health care needs. *The American Statistics Index* is the master index to all publications of the federal government that contain statistical information. It is accessible through ORBIT, DIALOG, BRS, and NTIS. Other selected statistical sources include *Statistical Abstract of the United States*, *A Guide to Sources of Statistical Information on Major Medical Topics*, *Standard Medical Almanac*, *Vital Statistics of the United States*, *Morbidity and Mortality Weekly Report*, *Surveillance Reports*, *Selected National Data Sources for Health Planners*, *Health: United States*, *Health Resources Statistics*, *Hospital Statistics*, *Reference Data on Profile of Medical Practice*, *Reference Data on Socioeconomic Issues in Health*, and *World Health Statistics Quarterly and Annual*.

IV. APPENDIX

A. DIRECTORY OF WASHINGTON HEALTH DATA SOURCES

U.S. Government

Department of Commerce
  Bureau of Census .... (202) 763-4040
  National Technical Information Service .... (703) 487-4600
Congress
  Congressional Budget Office .... (202) 226-2650
House
    Appropriations Subcommittee on HHS Appropriations .... (202) 225-3508
    Budget, Health Staff .... (202) 226-7115

Energy and Commerce, Subcommittee on Health and the
Environment ... (202) 225-0130
Ways and Means, Subcommittee on Health ... (202) 225-7785

Senate
Appropriations Subcommittee on HHS Appripriations ... (202) 224-7283
Budget, Health Staff ... (202) 224-7993
Finance, Subcommittee on Health ... (202) 224-4515
Labor and Human Resources ... (202) 224-3191

Office of Technology Assessment,
Health Program ... (202) 226-2070

Environmental Protection Agency ... (202) 382-4454

General Accounting Office
Human Resource Division ... (202) 275-5470

Department of Health and Human Services
Administration on Aging ... (202) 245-0641
Alcohol, Drug Abuse, and Mental Health ... (301) 443-4408
Assistant Secretary for Planning and Evaluation ... (202) 245-1858
Health Care Financing Administration ... (202) 245-6113
Health Resources and Services Administration ... (301) 443-2403
National Center for Health Services Research ... (301) 443-4100
National Center for Health Statistics ... (301) 436-8500
National Institutes of Health
Main number ... (301) 496-4000
National Library of Medicine ... (301)496-6095
National Institute of Mental Health ... (301) 443-4515
Office of Disease Prevention and Health Promotion ... (202) 245-7611
Office of Human Development Services ... (202) 472-7257
Social Security Administration ... (202) 673-5602

Department of Labor,
Bureau of Labor Statistics ... (202) 523-1221

Veterans Administration ... (202) 389-2525

Associations and Private Research Groups

American Association of Homes for the Aging ... (202) 296-5960
American Health Care Association ... (202) 833-2050
American Hospital Association ... (202) 638-1100
American Medical Association ... (202) 789-7400
American Nurses Association ... (202) 789-1800
American Public Health Association ... (202) 789-5660
Association of American Medical Colleges ... (202) 828-0455
Association of Schools of Public Health ... (202) 842-4668
Association of State and Territorial Health Officials Foundation ... (301) 933-7950
Federation of American Hospitals ... (202) 833-3090
Group Health Association of America ... (202) 737-4311
Health Industry Manufacturers Association ... (202) 452-8240

National Association of State Units on Aging
National Data Base on Aging ... (202) 424-9126
National Clearinghouse for Alcohol Information ... (301) 468-2600

National Council on the Aging ... (202) 479-1200

National Governor's Association
Health Policy Group ... (202) 624-5344

Pharmaceutical Manufacturers Association .... (202) 835-3400
Washington Group on Health.. (202) 547-6644

B. DIRECTORIES AND BIOGRAPHICAL SOURCES

American Men and Women of Science and its companion publication, *AMWS: The Medical and Health Sciences*, *American Medical Directory*, *Directory of Medical Specialists*, *American Dental Directory*, *Encyclopedia of Associations*, *Foundations Directory*, *Research Centers' Directory*, *Medical and Health Information Directory*, the *Health Care Directory*, *Scientific, Technical and Related Societies of the United States*, and the *AHA Guide to the Health field* are all useful sources of persons and organizations involved in health.

C. HISTORY SOURCES

Among the more valuable reference histories are Garrison's *An Introduction to the History of Medicine*, Dolan's *Nursing in Society: A Historical Perspective*, Thompson and Goldin's *The Hospital: A Social and Architectural History*, Weinberger's *History of Dentistry*, and Kremers' *History of Pharmacy*.

REFERENCES

1. **Runkel, P. J. and McGrath, J. E.,** *Research on Human Behavior: A Systematic Guide to Method,* Holt, Rinehart & Winston, New York, 1972.
2. **Durkheim, E.,** *The Rules of Sociological Method,* The Free Press, New York, 1938.
3. **Wallace, W. L.,** *The Logic of Science in Sociology,* Aldine Publishing, Chicago, 1971.
4. **Blumer, H.,** *Symbolic Interactionism,* Prentice-Hall, Englewood Cliffs, NJ, 1969.
5. **Popper, K. R.,** *The Logic of Scientific Discovery,* Harper & Row, New York, 1957.
6. **Popper, K. R.,** *Conjectures and Refutations: The Growth of Scientific Knowledge,* Harper & Row, New York, 1963.
7. **Feyerabend, P.,** *Against Method,* Verso, London, 1978.
8. **Kuhn, T. S.,** *The Structure of Scientific Revolutions,* 2nd ed., University of Chicago Press, Chicago, 1970.
9. **Kuhn, T. S.** *The Essential Tension: Selected Studies in Scientific Tradition and Change,* University of Chicago Press, Chicago, 1977.
10. **Fisher, R. A.,** *The Design of Experiments,* Oliver & Boyd, London, 1935.
11. **Campbell, D. T. and Stanley, J. C.,** *Experimental and Quasi-Experimental Designs for Research,* Rand McNally, Chicago, 1963.
12. **Blalock, H. M.,** *Theory Construction: From Verbal to Mathematical Formulations,* Prentice-Hall, Englewood Cliffs, NJ, 1969.
13. **Zetterberg, H. L.,** *On Theory and Verification in Sociology,* 3rd enlarged ed., Bedminster Press, Totowa, NJ, 1965.
14. **Glaser, B. G. and Strauss, A. L.,** *The Discovery of Grounded Theory: Strategies for Qualitative Research,* Aldine, Chicago, 1967.
15. **Pelto, P. J. and Pelto, G. H.,** *Anthropological Research: The Structure of Inquiry,* 2nd ed., Cambridge University Press, London, 1978.
16. **Lofland, J.,** *Analyzing Social Settings,* Wadsworth, Belmont, CA, 1971.
17. **Bogdan, R. and Taylor, S. J.,** *Introduction to Qualitative Research Methods,* John Wiley & Sons, New York, 1975.
18. **Fagothey, A.,** *Right and Reason: Ethics in Theory and Practice,* C. V. Mosby, St. Louis, 1976.
19. **Beauchamp, T. L. and Childress, J. F.,** *Principles of Biomedical Ethics,* 2nd ed., Oxford University Press, New York, 1983.
20. **Veatch, R. M.,** *A Theory of Medical Ethics,* Basic Books, New York, 1981.
21. **Berg, K. and Tranoy, K. E., Eds.,** *Research Ethics,* Alan R. Liss, New York, 1983.
22. **Myers, M. J.,** Ethics, in *Remington's Pharmaceutical Sciences,* 16th ed., Gennaro, A. R., Ed., Mack Publishing, Easton, PA, 1980, 19.
23. **Tranoy, K. E.,** Is there a universal research ethics?, in *Research Ethics,* Berg, K. and Tranoy, K. E., Eds., Alan R. Liss, New York, 1983, 3.

24. **Talalay, P., Ed.,** *Drugs in Our Society,* Johns Hopkins University Press, Baltimore, 1964.
25. **Campbell, A. V.,** *Moral Dilemmas in Medicine,* 2nd ed., Churchill Livingstone, Edinburgh, 1975.
26. **Cassileth, R. B., Luck, E. J., Miller, D. S., and Hurwitz, S.,** Attitudes toward clinical trials among patients and the public, *J. Am. Med. Assoc.,* 248, 968, 1982.
27. **Fletcher, J. C.,** The evolution of the ethics of informed consent, in *Research Ethics,* Berg, K. and Tranoy, K. E., Eds., Alan R. Liss, New York, 1983, 187.
28. **Gray, B. H.,** *Human Subjects in Medical Experimentation,* John Wiley & Sons, New York, 1975.
29. **Levine, R. J.,** *Ethics and Regulation of Clinical Research,* Urban and Schwarzenberg, Baltimore, 1981.
30. **Lebacqz, K.,** Controlled clinical trials: some ethical issues, *Controlled Clin. Trials,* 1, 29, 1980.
31. **Spodick, D. H.,** The randomized controlled clinical trial: scientific and ethical bases, *Am. J. Med.,* 73, 420, 1982.
32. **Zelen, M.,** A new design for randomized clinical trials, *N. Engl. J. Med.,* 300, 1242, 1979.
33. **Schafer, A.,** The ethics of the randomized clinical trial, *N. Engl. J. Med.,* 307, 719, 1982.
34. **Cranford, R. E. and Doudera, A. E.,** *Institutional Ethics Committees and Health Care Decision Making,* Health Administration Press, Ann Arbor, MI, 1984.
35. **Gray, B. H., Cooke, R. A., and Tannenbaum, A. S.,** Research involving human subjects, *Science,* 201, 1094, 1978.
36. **Boulding, K. E.,** Science: our common heritage, *Science,* 207, 831, 1980.
37. **Hempel, C. G.,** *Fundamentals of Concept Formation in Empirical Science,* University of Chicago Press, Chicago, 1952.
38. **Hempel, C. G.,** *Aspects of Scientific Explanation,* Free Press, New York, 1965.
39. **Reynolds, P. D.,** *A Primer in Theory Construction,* Bobbs-Merrill Company, Indianapolis, 1971.
40. **Hage, J.,** *Techniques and Problems of Theory Construction in Sociology,* John Wiley & Sons, New York, 1972.
41. **Freese, L.,** Formal theorizing, *Annu. Rev. Sociol.,* 6, 187, 1980.
42. **Beck, W. S.,** *Modern Science and the Nature of Life,* Harcourt, Brace, New York, 1957.

Chapter 3

RESEARCH DESIGN AND DATA COLLECTION METHODS

P. Hurd, E. Nnadi Okolo, A. G. Hartzema, C. O. Airhihenbuwa, and M. Wang

TABLE OF CONTENTS

* P. Hurd.
** E. Nnadi Okolo.
*** A. G. Hartzema.
† C. O. Airhihenbuwa.
‡ M. Wang.

I. RESEARCH DESIGNS

In discussing research design, the difference between an experimental approach and any other research method is important. Many projects are similar to a true experiment, but lack one or more crucial elements. In other situations, the nature of the research quesiton requires a nonexperimental approach because of such issues as cost, convenience, and available data. The interpretation of the results of a project will depend, in part, on the limitations of the actual design used to collect information.

Two of the fundamental components of an experimental approach are control and randomization; both are central in providing the researcher with an appropriate design and interpretable data.

Control is typically achieved when the experimenter is responsible for manipulating a particular aspect of the design. For example, three dosage levels of a drug are chosen by the experimenter and compared to a group receiving no medication. In this case the experimenter is ''in control'' of the medication that the subjects are receiving. No other medications are given during this session so that the patient responses can be easily compared across treatment conditions.

Another use of the term ''control'' is important in the design of true experiments; the ''control group'' refers to the subjects who serve as the comparison group for those receiving some sort of treatment. In a drug study, the controls would typically receive a placebo or no treatment at all. The ''control group'' helps the experimenter gain ''control'' of the treatment groups because it indicates what could have been expected to happen as a result of no intervention.

Using a control group, incidentally, acknowledges that the experimenter cannot control all the variables that may impact on the subjects in an experiment. Some aspects are beyond the control of the experimenter, are too difficult to control, or are not considered worth the effort to control. For example, in a drug study, personal attributes like the ability to ski, length of the nose, and depth of one's tan will vary across subjects, but will not be controlled by the experimenter. Using the control group helps the experimenter rule out these aspects of the subjects. The control group works because subjects have been randomly assigned to treatment and control conditions. Therefore, the distribution of these uncontrolled variables should be random. Their effects on experiments may add confusing "noise" to the results, but that interference will be distributed across conditions. Because individuals are assigned at random to a treatment or control group, whatever differences there are will be assigned without any plan or method, resulting in no systematic bias to distort the results. Of course, it is always possible that fate will result in some strange mix of variables that will distort the results: thus, the need for replication of a study is most important.

A. EXPERIMENTAL DESIGN

An experiment typically has both dependent and independent variables. Dependent variables are the outcome measures of an experiment. These would include the therapeutic effect of the treatment or the change in knowledge following an educational program. These measures are predicted to be "dependent" on the experimental condition to which the subject is assigned or the experimental treatment that the subject receives. Independent variables, on the other hand, are determined by the experimenter prior to the actual start of the project. The conditions of the experiment (drug one, drug two, placebo) or the treatment to which the subject is randomly assigned (education program, no education program) are examples of independent variables. The subject's "independent variable" status has been determined prior to the actual experiment.

Components of a 2 × 2 Design — In this type of design, each independent variable has two levels. In a drug study, one dimension would be drug or placebo. The second dimension could be based on some other intervention such as telling the individual to take a vitamin supplement at the same time as the medication or to take no vitamin supplement during the study. Subjects would be randomly assigned to one of four groups. Some would take both vitamin and drug, some would take placebo and no vitamin, some would take the vitamin and the placebo, and the fourth group would take only the drug. The control group in this study would be those taking the placebo without the vitamin.

Many experiments are a modification of a 2 × 2 design. For example, a 2 × 2 × 2 design would have three independent variables, each having two levels, and a 4 × 3 design would have two independent variables, one with four levels and one with three. In each case, subjects are randomly assigned to conditions, and there is an equal chance of being assigned to every one of the conditions in the experiment.

Occasionally, the research study will use characteristics of the subjects as one of the dimensions of the design of the project. In the drug study example, males and females may be randomly assigned to receive either drug or placebo. In this case, subjects cannot be randomly assigned to each of the four cells of the experiment because their characteristic of gender is unchangeable. This type of design is called a **mixed design.** Other examples of classifications that could be used in mixed designs would be age categories, personality dimensions, and categories based on some other assessment device (e.g., low, medium, or high grades in a class). The interpretation of mixed designs should differentiate between results from the randomly assigned conditions and those from the different classifications. For random assignment, one can typically speak of the variable causing something in the study. For instance, the drug caused the blood pressure to drop compared to the control group. When using nonmanipulated classifications, one can speak of an association between the variable and the outcome (older subjects responded slower than younger subjects to the medication), but should avoid interpreting the data as if this classification variable ''caused'' something. It is possible that some additional dimension that happens to vary with the age of the subjects in the study is actually the ''cause'' of the findings. If the older subjects responded differently to the drug because of some physiological change, then it is actually the physiological decline rather than age that would explain the results.

B. QUASI-EXPERIMENTAL DESIGNS

Quasi-experimental research designs have a very important role in research, especially in areas where the use of human subjects makes a strict experimental design impossible, unethical, or impractical. As the term ''quasi-experimental designs'' implies, these designs appear much like a true experiment. The exception would be in the lack of control compared to what one would normally expect to find in the experimental design. For example, some quasi-experimental projects have ''nonequivalent'' control groups. Rather than randomizing subjects to treatment conditions, a study may compare equivalent hospitals, one with a new treatment and one with the existing program. Since there is no control for the type of patient at the two hospitals and for the level of care at each hospital, biases may creep into the results. In other quasi-experimental designs, the control group may be missing, as in a posttest-only evaluation of a continuing education program. Neither a pretest of the participants nor a comparison with those who did not take the program is present, making the evaluation of the results somewhat questionable and the conclusions somewhat speculative.

This lack of experimental control helps to explain why quasi-experiments have a number of threats to internal validity. A high level of internal validity refers to the fact that the impact of the experimental variables on the subject was indeed the effect that the experimenter had intended. In the quasi-experimental design it is occasionally more difficult to determine if outside forces (those not controlled by the experimenter) are in any way accountable for the outcome of the experiment.

The passage of time can complicate an experiment and bring the internal validity of the study into question. One effect of time would be on the individual. He or she may change over time, affecting the outcome of the study. This change may go undetected or uncontrolled in quasi-experimental designs. These threats to internal validity may be because of the maturation of the subject, changes in immediate past history that would affect the outcome, or changes in mortality rates of subjects between conditions.

The selection of subjects can also influence the outcome of a study. One of the ''classic'' problems arises when the researcher chooses a group that has scored especially well or especially poorly on some sort of test. If these individuals are followed over time, there is a tendency for them to ''regress toward the mean'' of the group. One way to think of this is in terms of the ''margin for error'' of a test. If you select only the best and the worst students, some of those individuals will have been included in these extreme groups due to measuring error. When these individuals are tested again at the end of the program, there is a tendency for them to score more toward the mean rather than as a more extreme case. If you select a group of individuals who did poorly on their medication knowledge test and then test them after an educational program, some of these individuals will appear to do better because of the ''regression toward the mean''. One way to control for this problem is to randomly assign some of these subjects to a control group of individuals who have scored poorly on the test, but receive no intervention. This groups' ''regression toward the mean'' serves as a baseline of comparison for the treatment group.

Threats to internal validity are by no means limited only to quasi-experimental designs. Many of the across-cell threats to internal validity are problems for a variety of experimental and quasi-experimental designs. Diffusion is one of the threats that can be most problematic; subjects gain information about other conditions in the experiment and then use that information to change their own attitudes or behaviors. For example, subjects in a control group will talk to individuals in the treatment group and then decide that what the treatment group is doing would be a good idea. Imitation of subjects in a treatment group by subjects not in that condition is much like diffusion. In this case the behavior of another subject is copied, as in an antismoking program when one individual copies a behavior (such as saying "no") that has been taught to another individual. In this case the effect may or may not be long lasting. Rivalry between conditions also can have an unanticipated effect on a study when individuals in different groups realize that they are receiving different interventions from the researcher. In other situations there may be some attempt on the part of either the experimenter or the subject to compensate for the condition that they are in when they can determine that theirs is not the treatment condition of their choice.

C. *EX POST FACTO* RESEARCH

In *ex post facto* research, the researcher is unable to control independent variables because they have either already occurred or they cannot be controlled by nature. This form of research can be referred to as investigation after the fact, or after the fact research. In *ex post facto* research, direct control is not possible. The researcher is unable to experimentally control or randomly assign subjects, and this is the major difference between experimental design and *ex post facto* design.

Ex post facto research often lends itself to a self-selection form of sampling procedure for the study sample. The subjects are therefore studied because they possess the characteristics of interest to the research. Subjects with similar characteristics are therefore put in the same study group. This is unlike the other forms of sampling methods, where some form of random assignment is made. Suppose a researcher is interested in investigating the cause of a high incidence of deaths associated with liver cancer in a specific neighborhood. This form of research will be *ex post facto* in nature because the researcher has to investigate several causes, including life-styles, and attempt to find a common link between the victims. The research can at best only arrive at a probable cause since the researcher cannot control the independent variables by randomization or manipulations. The subjects either are already dead due to cancer of the liver or they already have cancer of the liver.

Perhaps the classical example of *ex post facto* research is one in which the investigator finds a high incidence of lung cancer, investigates several possible causes, and identifies cigarette smoking as the cause.

1. Disadvantages and Advantages

Ex post facto research is weaker than experimental design because of some inherent limitations. There are three major limitations in *ex post factor* research. *Ex post facto* research lacks control due to (1) inability to randomize and (2) inability to manipulate independent variables because of its retrospective nature. The third limitation (3) is the higher possibility of incorrect interpretation, thus being misleading compared to experimental designs. The results of *ex post facto* research can easily be misinterpreted if the research is conducted without hypotheses and predictions.

Although *ex post facto* research has some inherent weaknesses, it does have some merits. Some research problems can only be investigated as *ex post facto* because experimental design will not be possible. Several research problems in the areas of medical psychology, medical sociology, health education, and medical anthropology can only be investigated through *ex post facto* research. *Ex post facto* research findings have made significant contributions in health research. Many of these studies have linked life-style-related factors to various diseases and illnesses, e.g., smoking and lung cancer, drug- and alcohol-related congenital birth defects, and acquired immune deficiency syndrome (AIDS).

II. DATA COLLECTION METHODS

A. OBSERVATIONAL METHODS

Although all research involves observation of some sort, the observational method refers to studies that typically lack an effort on the part of the researcher to control the situation. Observational methods are differentiated from interviews and questionnaires because of the structure that is imposed by a written questionnaire or an interview with a specific format to follow. Naturally, there is some overlap of categories, as an unstructured interview would certainly fall into the observational methods category, as would a survey that was done to determine "what was going on" rather than to test a specific hypothesis. Case studies often will fall into the observational method category because they tend to be a report of what took place, sometimes with documentation, rather than an intervention on the part of those conducting the study. Participant observation would also fall into this category. Participant observers are individuals who immerse themselves in the situation that they are observing, often keeping a journal of events, and then report on the experience. An individual may join a street gang, ride with a police officer to observe behavior on the job, or feign an illness to explore how others will react.

One of the risks of this method is reactivity. Subjects who know that they are being observed will react differently than if they were in an unobserved situation. For example, workers who realize that an efficiency expert is in the hospital may structure their work so that they appear to be busy, but are actually working at less than full capacity. Consequently, the efficiency expert will set goals for the workers that will be easily achievable because they are below the normal level of operation of the workers. As another example, physicians who are aware of an effort to monitor the use of a certain medication may change their prescribing behavior during the time of the study, but then return to their preferred approach when data collection has concluded.

Observational methods need not be noticeable. Valuable research has been done by those using "unobtrusive" methods.* One of the best examples is based on the Chicago Museum of Science and Industry, which has a very interesting display of hatching eggs and newborn chicks. Because this display is so popular the floor tiles needed considerably more attention than other areas of the museum. One of the unobtrusive measures of interest in this display became the wear and tear on the floor in front of the exhibit. Other unobtrusive measures include observing from a distance, records that are already collected, placement of objects, or the time until a particular event occurs.

Demand characteristics are aspects of a study that raise the possibility that subjects are reacting to the demands of the researcher rather than acting as they would in an unobserved setting. The patients who report less pain after taking a medication because they know the researcher wants them to feel better or the group members who "stick to their guns" rather than change their opinion may be doing what they think will help the experimenter or acting as they think they are expected to behave. The experimenter may inadvertently communicate to the patient what is desirable behavior ("Your pain is better now, isn't it, Mr. B?"), or the subject may guess at what the appropriate behavior would be. In both cases, the validity of the study is affected and the research will be less likely to accurately predict what people would do in a different setting.

The case study is a type of observational study method that is seen in the health care setting, business schools, law schools, and other situations where the background and specific components of a situation are recorded. These cases are extremely valuable, but have the drawbacks of the absence of control groups and replications across subjects. A real value of case studies is their ability to illustrate a particular problem, principle, or decision process.

* Webb, E. J., Campbell, D. T., Schwartz, R. D., and Sechrest, L., *Unobtrusive Measures: Nonreactive Research in the Social Sciences,* Rand McNally, Chicago, 1966.

Case-control studies are a much more elaborate research method based on the area of epidemiology. In this approach a number of cases are chosen, such as all with the same disease state; controls are then selected who are similar to the cases except for the absence of the disease. Comparisons are then made between the two groups to identify risk factors that are higher for the cases than for the controls. While accurate selection of cases and controls is crucial to the validity of this design, conclusions are based on findings across a number of cases.

B. INTERVIEWS AND QUESTIONNAIRES/SURVEYS

When using an interview or questionnaire, the researcher wants to get an accurate picture of the content area of interest. Part of an accurate picture would be one that is unbiased and precise. An unbiased question would be one that does not suggest that one type of response is preferred. Avoid such biased questions as agree-disagree answers to ''No one in their right mind would support socialized medicine'' or ''Every red-blooded American man has the right to own as many guns as he wants''. These questions should be rewritten to avoid disclosing the opinion of the writer. ''Precise'' has to do with the precision of the measurement. To measure the height of patients one would not use an elastic measuring tape because the determination of height could vary. To measure blood pressure one would not round off to the closest ten (120/80 or 120/90 instead of 120/85).

Interviews can be based on a very strict series of questions or on a less formalized approach. Some interviews use an extremely structured format, following a specific series of questions with specific responses to those questions. This type of interview often uses an elaborate schedule of questions and is highly standardized. Each individual should be asked identical questions in the exact same order. This type of interview is called a scheduled interview. Some interviews are less closely controlled, perhaps defining certain topics or areas that must be covered in the interview. These are called semistructured interviews. Some interviews are unscheduled and follow different directions with different individuals. This approach might be most appropriate when a researcher is seeking a very broad understanding of the individuals who are interviewed or is at a very early stage in the development of the interview.

1. Guildelines for Written Questionnaire Construction

When creating a form for data collection there are a number of errors that are to be avoided. Each error is based on the tendency to think of these forms from the researcher's perspective rather than from that of the person filling out the survey. Many individuals who are completing a questionnaire find the activity boring rather than interesting, time-consuming rather than self-fulfilling, and dull rather than stimulating. Some individuals, such as hospital or pharmacy administrators, are constantly asked to fill out questionnaires, often with little or no reward. These individuals have many reasons to throw the paper in the trash rather than complete the form and send it back to the anxious researcher.

Ask yourself, ''Why do I want to know this?'' for each question that you ask. You *must* have a good reason for collecting the data. Without this reason you will simply collect and code the data, probably never analyzing or reporting the results of your findings for that question. If you find yourself saying ''Maybe I'll want to look at this later'' you should seriously consider dropping the question. A questionnaire filled with these unnecessary questions will be longer (reducing the response rate in most situations), may ask a question that the respondent does not want to answer, or may affect the responses to other questions as a result of providing this information. Only ask questions that you have a specific need to have answered.

Ask the respondent the demographics (age, sex, address, etc.) at the end. This information is uninteresting to the respondent and requires the repetition of facts that have been given many times. The questions that are central to the questionnaire should be asked early in the questionnaire.

Ask especially sensitive questions in the middle of the survey. The participant needs to become involved in the procedure and committed to the project before a very personal question is tackled. Even if it is only through the written questions in a mailed survey, there is a relationship that develops between the individual and the researcher. This "relationship" can help the researcher to increase response rate and validity if considered in the development of the questionnaire.

Keep the questions interesting. If the material is boring, repetitious, or tedious, the individual will lose interest in answering. Time spent on the flow of the questionnaire and the construction of individual questions can lead to a better data collection instrument.

Keep the questionnaire easy. The more work the researcher spends on developing easy responses, the more pleasant the experience for the respondent. In multiple-choice-type questions, use options that will account for most of the possibilities. It is easier to check a particular response than to decide that none apply. Leave an option for a response other than those listed when possible. It is quite frustrating to be unable to indicate your response because your answer is not listed among the options. For example, if you work full time in the community and part time in a hospital, you would have trouble with:

Indicate your place of employment:
________ community
________ hospital

Another mistake to avoid is the overlapping of categories. Make sure that only one response is appropriate for each individual.

What is your current annual salary?
________ Under $20,000
________ $20,000 to $25,000
________ $25,000 to $30,000
________ $30,000 to $35,000
________ $35,000 to $45,000
________ $45,000 to $55,000
________ more than $55,000

If you make exactly $25,000 or $35,000 or $45,000 you will not know exactly how to answer the questionnaire. Another mistake in this questionnaire is the use of unequal intervals at the higher salaries. You have changed the response from interval data to ordinal data, greatly complicating the statistics and potentially confusing the respondent.

The salary question above does attempt to cover all options. Another error in question construction is to exclude responses that an individual would like to make. A question like "Are you now taking any prescription medicines on a regular basis?" may present a problem for an individual taking a prescription medication for pain on an as-needed based, especially if the individual is trying to avoid using the medicine. The inclination might be to answer with a "no" even though the researcher might intend for this question to be answered "yes" by this type of individual.

Another problem in question design is the use of overlapping questions — for example, "Do you take any over-the-counter medications or vitamins?" The question itself is less of a problem than the interpretation by the researcher. This question will NOT tell you if someone takes ONLY vitamins and it will NOT tell you if they take over-the-counter (OTC) medicines other than vitamins. People answering "yes" to this question will be in three categories: vitamins only, OTCs other than vitamins only, and both. For some research questions this is acceptable, but often this overlap type of question will give the researcher difficulty in the analysis. This style of question is almost always best to avoid.

Another common error in questions that are designed to meet a common format is the mismatch of the question and the response categories. If you are using 1 = completely satisfied and 7 = completely dissatisfied, the following sequence would present a problem for the subject:

The medication has helped my problem.
The physician took time to explain the medicine to me.
The patient should be counseled about his/her medications whenever possible.

The third question does not fit well on the satisfaction dimension. The first two questions could be answered on this dimension, although the subject might find it easier to answer if the question was "strongly agree to strongly disagree".

Another aspect to consider is the "punishment" of the conscientious responder. Questions will sometimes ask for more information if a certain response is made.

"Do you have a unit dose distribution system in your hospital?
YES
NO . . . if "NO" please describe the system that you use."

In some situations the individual is punished for responding in a certain way in the sense that he or she must explain the answer that has been chosen for question after question. Although this style of questioning is most effective when an individual occasionally can explain a personal answer to the question, the researcher should avoid designing a series of questions where a certain response will mean more writing on the part of the subject. The easier the questionnaire is to answer, the better the chances of getting a high rate of return on the survey.

Questions that are ambiguous are also to be avoided on a questionnaire. Not only will the subject find them difficult to answer, but the researcher will also find the explanation of the data virtually impossible because no one knows how the subjects interpreted the question. For example, in a question about the various ways that patients might receive medications, you could ask:

Do you use mail order? YES NO

Would the researcher mean mail order for prescription drugs, or does this include vitamins? How about OTC products? Would some subjects answer "yes" if they ordered shirts and pants from a mail-order catalogue? If this question is used, the researcher will receive answers; the data, however, will be useless.

Avoid leading or threatening questions. The discussion about bias in questions included examples of leading questions. These questions cue the respondent as to the answer that the experimenter thinks is best. "The government should not waste its money on excessive military spending" and "The government should provide for an adequate military defense" could both be answered in the affirmative, although the second question could be taken as support for military spending while the first question could be interpreted as support for less military spending. Leading and threatening questions tend to distort the responses that are provided. Threatening areas would include income, personal weaknesses, substance abuse, and the evaluation of others with whom one works. This type of question can be asked, but the researcher needs to use special care with this type of data collection to avoid a low level of response, responses that are inaccurate, and questions that are simply skipped by the subject.

Try to phrase questions in a positive way, avoiding the use of works like "no", "not", and "never". Which sounds easier to answer: "I have no trouble taking my medications" or "I take my medications without any trouble"? The second phrasing of the question avoids the negative. The word "no" is sometimes missed visually — the subject reads the sentence without this word and gives the opposite answer. Those taking multiple-choice tests are sometimes victimized by this tendency. Positive statements tend to yield more usable data.

2. Advantages of Survey Research

The survey approach offers the advantage of wide scope and the study of large populations by selecting and studying samples chosen from the populations to discover the relative incidence, distribution, and interrelationships of sociological and psychological variables. Survey research focuses on people, their beliefs, opinions, attitudes, motivations, and behavior, and is therefore well suited for studies dealing with health behavior, attitudes, and utilization of health services.

Dean et al.[1] indicated the rationale of survey research as follows:

> Survey method's major advantage is that it gives to social scientists a quantitative method for establishing relationships and for generalizing about populations. A systematically selected sample of individuals is exposed to a fixed set of questions; then their reactions to those questions are systematically classified so that quantitative comparison can be made....This standardization also creates one of the survey method's greatest weaknesses; it exposes each respondent to individual questions and classifies his responses into a few simple types regardless of the distinct qualities of each response.

Babbie[2] cites five scientific merits of the survey research method:

1. Survey research is logical. The format of survey research often permits the rigorous, step-by-step development and testing of such logical explanation.
2. Survey research is deterministic. Whenever the survey researcher attempts to explain the reasons for and sources of observed events, characteristics, and correlations, he must assume a deterministic posture. Where the survey format permits a clear and rigorous elaboration of a logical model, this clarifies the deterministic system of cause and effect.
3. Survey research is general. Sample surveys are almost never conducted for purposes of describing the particular sample under study. Rather, they are conducted for purposes of understanding the larger population from which the sample was initially selected.
4. Survey research is parsimonious. Since the survey format lends itself to the collection of many variables that can be quantified and processed by machine, the survey analyst can construct a variety of explanatory models and then select the one best suited to his aims.
5. Survey research is specific. Scientific research aims at ever more sophisticated and more useful conceptualization and measurements, but at every step along the way, methods used must be made specific. Survey research by its very nature lends itself readily to this.

Survey research investigates variables which are sociological facts (i.e., sex, income, religious affiliation, socioeconomic status, education, race, etc.) or psychological, which includes opinions, attitudes, and behaviors.

The survey design can utilize cross-sectional format or longitudinal format. Cross-sectional format means that data collection occurred at one point in time from the target population, while in longitudinal format data collection from the sample occurred at different times during the course of the study. The major drawback of cross-sectional measurement, unlike longitudinal measurement, is that of indicating causality, since a group's activities and its inherent changes are not monitored over time as in a longitudinal survey research study.

REFERENCES — SECTION II.B.2

1. **Dean, J. P., Elchorn, R. L., and Dean, L. R.,** The survey, *Introduction to Social Research,* Doby, J. T., Ed., Appleton-Century-Crofts, New York, 1967, 147.
2. **Babbie, Earl R.,** *Survey Research Methods,* Wadsworth Publishing, Belmont, CA, 1973, 47.

3. A Guide to Training Research Interviewers

INTRODUCTION

My name is______________________________. I am here to help with the training of interviewers for the ______________________________ project. At the end of training you may be required to take an examination followed by an interview. Those who are successful with both the exam and interview will then receive further instructions and later perform the actual interview, in the areas which have been selected.

Please, if you have any questions, feel free to ask.

Please, if you are not selected, it will be because other individuals performed better than you did. So it is *not* because anyone does not like you. So, if you want to be selected, pay attention and perform satisfactorily.

There are four steps in the training process:

Step I: Questionnaire administration
Step II: Evaluation (examination) of participants
Step III: Practice/demonstration on questionnaire administration and pretesting of questionnaire
Step IV: Post-pretest

QUESTIONNAIRE ADMINISTRATION TRAINING OUTLINE

Step I — Questionnaire Administration Procedure

1. PURPOSE
2. THE SURVEY QUESTIONNAIRE (MAJOR TYPES)
   — Key advantages and disadvantages of each
3. SELF-ADMINISTERED QUESTIONNAIRE
4. PERSONAL INTERVIEW (FOR THOSE WHO CAN'T WRITE AND READ)
   — Bias
   — Misunderstandings, clarifications, etc.,
   — Interruptions

5. IMPORTANCE OF CONFIDENTIALITY
6. IMPORTANCE OF BEING COURTEOUS
7. DEALING WITH POTENTIAL PROBLEMS
8. CRITERIA FOR RESPONDENTS
9. PROCEDURE FOR QUESTIONNAIRE ADMINISTRATION
   — Prior
   — During
   — After
10. HANDLING OF COMPLETED QUESTIONNAIRE
   — Indicate where to return completed questionnaire.

Step II — Evaluation

1. EVALUATION
2. FINAL SELECTION

Step III — Practice Exercise

1. PRACTICE EXERCISE
2. PRETEST QUESTIONNAIRE

Step IV — Post-Pretest

1. DISCUSS AND DEAL WITH PROBLEMS ENCOUNTERED DURING PRETEST

STEP I

MAIN PURPOSE:

A questionnaire in general is used as an instrument or as a means of collecting information, often for research purposes. This form of gathering information is sometimes called a survey.

The purpose of training is to give the interviewers adequate information on proper techniques for administering the questionnaire, thereby making the interviewers knowledgeable in appropriate questionnaire administration. Training is very important and should be taken very seriously. Administering questionnaire the wrong way will reduce the usefulness of the information or even make the information useless.

Purpose of this Research Study: We hope to obtain information regarding ________. (Indicate the purpose of the study, not anticipated results. Also indicate how the results can be useful in the area of health.)

MAJOR TYPES OF SURVEYS:

There are different ways of doing surveys, such as:

(1) Mail survey (mailing questionnaire to participants)
   Major disadvantages
   (a) Poor response rate
   (b) Loss of control — the intended respondent in the household may not be the person who completed the survey
   (c) Can only be used for individuals who can read and write

Major advantages
(a) Cheaper than personal interview
(b) Reduced possibility of interviewer bias

(2) Telephone survey
Major disadvantages
(a) Only people with at least one telephone can be reached
(b) May not reach individuals with unlisted phone numbers
Major advantages
(a) Saves travel time
(b) Better control than mail survey

(3) Personal interview survey
This can be self-administered or administered by an interviewer.
Major disadvantages
(a) Expensive
(b) Time-consuming, especially in travel time
(c) Possibility of interviewer bias especially when not properly trained (bias will be discussed later)
Major advantages
(a) Maintain good rapport (relationship) with respondent
(b) Can explain the question (not answer it)
(c) Can deal with some problems better, eg., omission, misunderstandings, etc.
(d) Can reach many people including those who cannot read and write, such as children and illiterate adults

This training guide will therefore focus on self-administered questionnaires and personal interviews.

Self-Administered Questionnaires:

Self-administered questionnaires are used where the respondent can write and read. The respondent is therefore given the questionnaire and asked to complete it without help from anyone.

During self-administration the respondent can ask for clarification, in which case the person giving the survey can explain the question, but not answer it.

Personal Interview:

Where possible, make an appointment with a respondent for a suitable time to be interviewed. This is used where the respondent cannot read and write. It is also used for children and individuals with certain types of handicap, e.g., the blind.

During a personal interview the interviewer reads out the instructions and then proceeds with the questions in the questionnaire. The interviewer reads the questions, followed by the responses, and the respondent chooses the best response applicable to him. This procedure is continued until all questions are answered.

Later we will discuss the procedure for proper administration of a questionnaire. We will discuss what should be done before, during, and after questionnaire administration.

Bias:

Now let us discuss the issue of bias. Bias is introduced when an interviewer influences the responses given by a respondent. There are several ways of introducing bias:

(1) When the interviewer gives the responses by telling the respondent what to answer
(2) By making it difficult for the respondent to tell the truth by verbal or nonverbal (facial) expression

  (a) By giving any impression of a verbal or nonverbal (facial) expression
  (b) By asking a question in an embarrassing way

(3) Arguing with the respondent about his/her response
(4) Indicating the type of result you expect or want
(5) Changing the response
(6) Discussing the responses a respondent should give

Biasing a questionnaire is a serious matter because it makes the true results of the survey difficult to see. This is why it is very important that you do not give the responses or even suggest in any way the responses that a respondent should choose.

Misunderstandings/Clarifications

If a respondent does not understand the question, the question should be repeated. Do not change the meaning of the question. Simply repeat the question using an appropriate tone of voice, after which you will repeat the responses from which the respondent can choose. The respondent should choose only one best response unless otherwise specified.

Interruptions

Interruptions should be avoided. Therefore, before administering the questionnaire, find a private place where you will not be interrupted until the questionnaire is completed. Also prior to the questionnaire administration explain the importance of completing the questionnaire without interruption.

However, should interruption occur, especially if it is for a few minutes, continue the questionnaire after the interruption, but make a note of it on the questionnaire. Should the interruption involve several hours, end the interview and also make a note of it, but do not count it as completed. Put the questionnaire aside as incomplete and write "incomplete" on the front page of the questionnaire.

Importance of Confidentiality

The importance of confidentiality cannot be overemphasized. The responses given by a subject should never be discussed with anyone. All information given as a result of the interview must be kept confidential. To help maintain this confidentiality do not (never) write anyone's name on the questionnaire.

Always remember that the respondents are giving you the privilege of providing you with information and you should always respect that.

Importance of Being Courteous

You should be courteous at all times. You should never insult a respondent for any reason, and under no circumstances should you fight with them. Remember that you are a professional; therefore, always exhibit professionalism.

Always remember that you are representing a lot of people, not just yourself. You are indirectly representing the people involved with this study, so be as respectful as you can.

Dealing with Possible Problems

Sometimes problems are encountered in questionnaire administration. Some of the problems may include:

(1) If there is no adult (or no one) at home: If there is no one at home, come back another time. If there is someone at home, tell them who you are if the family does not know you, and let them know you will come back.
(2) If the respondent refuses to be interviewed: (this will be discussed later under Questionnaire Administration)

(3) Making sure that the questionnaire is entirely completed in case of self-administered questionnaire: Courteously show the respondent the questions he/she has missed and ask him/her to complete them. An incomplete questionnaire is not very useful.

(4) The respondent questions your integrity: Let him/her know that you were specially trained for this project and that you are therefore administering the survey on behalf of the project.

(5) Any other problems: Handle them accordingly and professionally. Always be professional. Remember: you are a part of a special project.

(6) If you get sick: Please contact______________________ or _________________ as soon as possible

Criteria for Respondent:

Discuss the criteria for the survey participants based on the study. Also give examples of those who are not eligible.

PROCEDURE FOR QUESTIONNAIRE ADMINISTRATION

Before (Preparing for Questionnaire Administration)

(1) Get all the materials you will need ready.
— pencils, pens, etc.

(2) Take about the number of questionnaires you want to complete for the day. After the first day you will have an idea of how many you can complete in a day.

(3) Take a few more just in case there is a mistake or a problem and you need another one.

(4) Always write directly on the questionnaire; do not record the interview on anything else.

When You Arrive at the Administration Site

(1) Always maintain a good impression.

(2) Be courteous (the importance of this has already been discussed).

(3) Introduce yourself (if you are not known to the family).
— Make sure the individual to be interviewed meets criteria.

(4) Explain the purpose of the survey.
— I, (name), represent the ______________________________

(5) Inform respondent that all information is kept confidential and his/her responses cannot be identified because his/her name is not on the questionnaire.

(6) Ask respondent to answer as truthfully as possible.
— Telling us the truth will only help. Not telling the truth will mean a lot of waste and will not be beneficial.
— There is no right or wrong answer; it is not an exam. The truth is the right answer.

(7) Then ask him/her to participate.

(8) Also tell him/her that it will not take much time.

(9) If they refuse

Explain the importance again, that confidentiality will be maintained and that it won't take much time. If he/she still refuses, then leave him/her alone. Make a note on a piece of paper, but do not write the person's name.

If they agree to be interviewed

— If the person can read and write, give him/her the questionnaire to fill out while you wait. When he/she finishes, check to be sure that it has been completed appropriately and completely.

— Do not give questionnaire away for respondents to complete later and return to you because you then lose control of respondent. If respondent cannot read and write then administer the questionnaire yourself.

Prior to administration find a private place

This is important to maintain confidentiality and avoid interruptions. If the respondent wants to know why it should be private, tell them this is for confidentiality and that they may find some questions personal.

During the Interview

(1) You must remain neutral to avoid bias (which has been discussed).
(2) Do not give any impression that any one response is not good.
 * Remember not to introduce bias, as discussed before.
(3) Be sure to tell respondent it is his/her response that is important, not what anyone thinks.
(4) Interview only one person at a time.
 — The questions should not be discussed in a group even though you may find the respondents as a group.
 — When you find them as a group, interview them separately one after another.
 — They can discuss the survey if they want to only after you have finished all the interviews, but never discuss the responses.
 — If you find those who can read and write as a group, give them the questionnaire separately; let them complete it alone separately and then collect the questionnaire when they finish. They should not discuss the questionnaire whilethey are answering.
 — The reason is to avoid anyone influencing someone else's response.
(5) Ask the questions and put a check, a mark (√), beside each appropriate response WITH A PEN. Do not change any question or response. Do not change the order of the questions.
(6) Check only one most appropriate response.
(7) Do not rush the interview.
 — Always remember that it is their opinion that is important.
(8) You may repeat the question or explain it without changing the meaning, but cannot answer it or tell the respondent what to answer.
(9) Make sure all questions are answered.
 — Put the completed questionnaires in a separate place from blank ones.
(10) The interview must be done on the questionnaire.

REMEMBER THE PROCEDURE

(1) Ask the question exactly as it is in the questionnaire.
(2) Then read the responses.
(3) Respondent will choose one best response and you will put a check mark (√) beside the response; then proceed to the next item until the questionnaire is completed.

After (at the End of the Interview)

When you have completed the questionnaire, check to be sure that you have not missed any questions.

Thank the respondent for his/her response. Then say goodbye and proceed to the next respondent.

Handling of Completed Questionnaire

All completed questionnaires and the return check list (to be given out) should be returned to ______________________________ immediately upon return from your assignment.

While on your assignment always separate completed surveys from the blank ones.

STEP II — EVALUATION

I. Introduction

A. Yesterday you had the first phase of your training as a research interviewer.

B. The process of training is a long one and will continue until you go to your assigned areas to conduct the actual interviews.

C. Today we will begin another phase of the training process, which is evaluation. The evaluation consists of an examination which you will take as soon as possible and the interview which will follow later.

II. Exam

In the exam you will be tested on the information (lecture) you received and the ability to follow instructions and work independently. So read the instructions and answer accordingly. The exam should take approximately 45 minutes.

III. Exam Procedure

A. The exam must be done individually without help from anyone.

B. Cheating will not be tolerated and will automatically disqualify you as an interviewer.

C. No questions or talking during the exam.

D. Use a pen.

E. When you finish submit (hand in) your completed examination to an investigator and leave.

IV. Grading

The exams will be graded as quickly as possible. Successful candidates will then proceed to the next phase of the training process, which is the interview.

Grading Priority

A. Following instructions

B. Total points

V. Results

VI. Interview

— Emphasize importance of following instructions

— Knowing the area for interview well

— If any suitable alternative interviewer is available

VII. Interviewers' Final Selection

— Alternative interviewers identified

PROJECT INTERVIEWEE
SAMPLE EXAMINATION

TOTAL — points

NAME: ______________________

Answer all questions on this paper.

1. In one short sentence, what is the KEY purpose of this research project? (4 points)

____

For questions 2 to 8 (7 points) answer True or False. For False put W in front of the question in the space provided. For True put X.

____ 2. Introducing bias is bad when it is a big bias, but a small bias is not bad.

____ 3. It is all right to answer or respond for a respondent if the respondent does not know what to respond.

____ 4. You should do personal interviews individually, that is, one person at a time, even if you find the respondents in a group.

____ 5. If you find someone who meets the criteria as a good candidate for an interview but they refuse to be interviewed, you should force them to be interviewed.

____ 6. Confidentiality is important, but you can discuss the responses with a close friend you trust.

____ 7. If you are interrupted for a short time during the interview, you can continue the interview when the individual returns.

____ 8. Being courteous is important, but if a respondent insults you you should insult them back, just a little.

9. Who should you submit completed surveys to immediately upon return? (1 point)

____

____

10. Name the two people who you may call if you encounter any problem.

____

____

11. List three major types of surveys. (6 points)

____

____

____

STEP III — PRACTICE EXERCISE

Demo/Practice Outline

Congratulations for successfully completing your exam. Today we will continue with your training process. This is the last phase of the training.

OUTLINE

1. Importance of following instructions
2. Returning of completed questionnaire
3. Practice exercise (role playing)
   Major objective of Phase III training: to practice administration of questionnaire so that you will be very familiar with the questionnaire before administering it. We will demonstrate to you how to properly administer a questionnaire.
4. Questionnaire administration exercise
   a. Read the instructions.
   b. Discuss the proper way of putting a (√) mark.
   c. Also explain to those who will self-administer the questionnaire the appropriate way of putting a check (√) mark. Examples of right and wrong ways of putting a check mark are given in Table 1.
5. Demonstration — use the questionnaire for the study to show interviewers appropriate and inappropriate techniques.
6. Practice by interviewers
   Instructions: (a) Write your name and the word "practice" in front of questionnaire.
   (b) Then form pairs (twos).
   (c) One person will act as the interviewer.
   (d) The other person will act as the interviewee. (Act as you feel that an interviewee will act during the interview.)
   (e) After the first exercise, change partners such that the interviewee in the previous exercise will have the opportunity to become an interviewer.
   — Submit your demonstration questionnaire to the instructor after the exercise.
7. Review the demonstration questionnaire individually.
8. Correct any problems individually.
9. What to do if questionnaire is stolen or destroyed:
   — Contact ______________________________ immediately for replacement.
10. Review questionnaire return check list.
11. Brief interviewers on pretest.
    1. When pretest is due
    2. Where to return pretest for review
12. Pass out questionnaires for pretesting.
    The interviewers should then administer the questionnaire to the sample selected for pretesting. After the pretest, schedule time with interviewers for Step IV, the final training step, to discuss the pretesting.

TABLE 1

Right & Wrong Ways of Putting a Check (√) Mark on Questionnaire

How old are you?
_______ Less than 21 yrs.
_______ 21 — 39 yrs.
_______ 40 — 50 yrs.
_______ More than 50 yrs.

WRONG

1. How old are you?
___√___ Less than 21 yrs.
_______ 21 — 39 yrs.
_______ 40 — 50 yrs.
_______ More than 50 yrs.

2. How old are you?
___√___ Less than 21 yrs.
_______ 21 — 39 yrs.
_______ 40 — 50 yrs.
_______ More than 50 yrs.

3. How old are you?
___√___ Less than 21 yrs.
_______ 21 — 39 yrs.
___√___ 40 — 50 yrs.
_______ More than 50 yrs.

RIGHT

4. How old are you?
_______ Less than 21 yrs.
___√___ 21 — 39 yrs.
_______ 40 — 50 yrs.
_______ More than 50 yrs.

5. How old are you?
___√___ Less than 21 yrs.
_______ 21 — 39 yrs.
_______ 40 — 50 yrs.
_______ More than 50 yrs.

6. How old are you?
_______ Less than 21 yrs.
_______ 21 — 39 yrs.
_______ 40 — 50 yrs.
___√___ More than 50 yrs.

On Table 1, the check marks on items #1 and #2 are wrong because the marks are ambiguous. It is difficult to determine if the respondent's age is less than 21 years or between 21 and 39 years. Item #3 is wrong for two reasons — first, due to ambiguity, and second, due to multiple check marks. These therefore may render this particular item on the questionnaire invalid. Items #4 to #6 are correctly marked because 1) they have no ambiguities and, therefore, there is no chance of misinterpretation, and 2) only one check mark has been made for each item.

RESEARCH PROJECT

QUESTIONNAIRE RETURN CHECK LIST

INTERVIEWER'S NAME: ______________________

Area Interviewed: ______________________

# of Questionnaires Received: ____________ Signature of Interviewer: ____________

Name of Facilitator: ______________________

# of Questionnaires Returned: ______________________

I hereby certify that I personally administered the questionnaire (except the self-administered ones) using the appropriate procedure taught during training.

Signature: ____________ Date: ____________

Number of individuals who refused to be interviewed ____________

# of Males: ____________ Approx. Ages ____________

# of Females: ____________ Approx. Ages ____________

Step IV — Post-Pretest Outline

1. Collect the pretest questionnaires after review and mark them as pretests.
2. Review each questionnaire individually to be sure that it has been properly completed.
3. Discuss problems encountered during the questionnaire administration in pretest.
4. Answer questions interviewers have.
5. Discuss any problem with individuals involved.
6. Make changes on the questionnaire if necessary based on difficulties encountered during pretest.
7. Have final briefing on questionnaire administration.
8. Develop code list with appropriate individual.
9. Distribute questionnaire to be administered and questionnaire return check list to interviewers.

Telephone Interviews

The major difference between personal interviews and telephone interviews is that whereas personal interviews are conducted face to face, telephone interviews are done over the phone. In telephone interviews, the interviewer has no knowledge of what the interviewee is doing and of any distractions during the interview process. The respondent may be getting help from another person without the interviewer's knowledge. However, telephone interviews are useful especially where a very large sample, covering several geographic areas, is involved in the study.

The steps for training a telephone interviewer and conducting telephone interviews are similar to those of personal interviews, which have already been discussed. The procedures are similar. The difference is that the interaction between the interviewer and interviewee is occurring over the telephone.

4. Examples of Research Questionnaires

a. The Health-Seeking Behavior Survey Instrument — Nigeria Study

The questionnaire was carefully constructed especially for the Nigerian population. Every effort was made to assure that the possible fixed responses were exhaustive and mutually exclusive. The questionnaire was developed in consideration of, but not limited to, the following factors:

1. Length of questions
2. Unidimensionality of the questions
3. Specificity of the questions
4. Culturally, socially acceptable questions
5. Clarity of the questions
6. Readability

HEALTH-SEEKING BEHAVIOR SURVEY INSTRUCTIONS

Please read the following very carefully. Then put a check (√) mark beside the answer that is most applicable for you. Choose only one most correct answer to each question.

Please answer as accurately as possible. These are no right or wrong answers. DO NOT write your name on this paper, because we do not want to know the particular person who answers the questions. Please answer all the questions.

Thank you very much for answering the questions in this form. Your cooperation is highly appreciated. Please ask questions if you do not understand something. Again, thank you for participating.

Immediate family = parents, sisters, brothers, wife/husband, and your children.

Please check (√) only one answer for each question.

1. Are you
   _______ Male
   _______ Female

2. What is your religion?
   _______ Catholic
   _______ Protestant
   _______ Muslem
   _______ Traditional worshipper
   _______ None of the above (specify; i.e., write in name) ____________________

3. If you have ever used native medicine when sick, in general, how satisfied were you?
   _______ Very satisfied
   _______ Satisfied
   _______ Dissatisfied
   _______ Very dissatisfied

4. If you have children, would you say that most of your children are
   Less than 10 years old
   _______ 10 years old to 20 years old
   _______ 21 years or more
   ____

5. How close do you live to a native doctor?
   Very close
   _______ Close
   _______ Not close
   ____

6. What is the (1st) first most important thing that helps you to decide where to go to for health problems?
   _______ Cost
   _______ Seriousness of the problem
   _______ Effectiveness (possibility of cure)
   _______ Distance to be traveled
   _______ Cause of the illness
   _______ Other (specify)____________________

7. What is the (2nd) second important thing that helps you to decide where to go to for health problems?
   _______ Cost
   _______ Seriousness of the health problem
   _______ Effectiveness (possibility of cure)
   _______ Distance to be traveled
   _______ Cause of the illness
   _______ Other (specify) ____________________

8. What is the (3rd) third important thing that helps you to decide where to go to for health problems?

_______ Cost
_______ Seriousness of the health problem
_______ Effectiveness (possibility of cure)
_______ Distance to be traveled
_______ Cause of the illness
_______ Other (specify) ______________________________

9. What is the (4th) fourth important thing that helps you to decide where to go for health problems?

_______ Cost
_______ Seriousness of the health problem
_______ Effectiveness (possibility of cure)
_______ Distance to be traveled
_______ Cause of the illness
_______ Other (specify) ______________________________

10. How often have you used English medicine when sick?

_______ Always
_______ Sometime
_______ Rarely
_______ Never

11. How close do you live to a hospital or clinic?

_______ Very close
_______ Close
_______ Not close

12. Are you working in a hospital, clinic, or maternity home?

_______ No
_______ Yes

13. Are any of your close relatives suffering from a long illness?

_______ No
_______ Yes

14. If you are working for the government, what will you say is your salary level now?

_______ Level 1 to 3
_______ Level 4 to 6
_______ Level 7 to 9
_______ Level 10 to 12
_______ Level 13 to 15
_______ Level 16 to 17

15. What is your occupation?
    _______ Civil Service (not teaching) = Ministry of _______________
    (write in)
    _______ Teaching
    _______ Hospital related work (e.g., nursing, hospital lab work, midwifery)
    _______ Housewife (check this only if not working)
    _______ Native doctor
    _______ Farming
    _______ Trading (businessman and woman)
    _______ None of the above (write in here) _______________

16. Would you say the head of your immediate family is a
    _______ Man
    _______ Woman

17. Would you say that you prefer
    _______ Native way of dressing
    _______ English way of dressing

18. Are any of your close relatives working in a hospital, clinic, or maternity home?
    _______ No
    _______ Yes

19. What is your tribe?
    _______ Hausa
    _______ Ibo
    _______ Yoruba

20. Whenever you get sick who do you go to first?
    _______ Native doctor
    _______ Parents
    _______ Your husband
    _______ Someone who works in the hospital
    _______ Other (who?) _______________

21. Would you say that, in general, members of your immediate family are usually
    _______ Very healthy
    _______ Healthy
    _______ Sick
    _______ Very sick

22. If you have ever used English medicine when sick, in general, how satisfied were you?
    _______ Very satisfied
    _______ Satisfied
    _______ Dissatisfied
    _______ Very dissatisfied

23. How much do you like going to the hospital or clinic when you are sick?
_______ I like it very much
_______ I like it a little
_______ I dislike it a little
_______ I dislike it very much

24. Where are you living now?
_______ City or township
_______ Village

25. Where will you go first for the following?
Diarrhea or dysentery
_______ Native doctors _______ Hospital _______ Other (specify) __________
Malaria
_______ Native doctors _______ Hospital _______ Other (specify) __________
Worms
_______ Native doctors _______ Hospital _______ Other (specify) __________
Convulsions (fits)
_______ Native doctors _______ Hospital _______ Other (specify) __________
High fever
_______ Native doctors _______ Hospital _______ Other (specify) __________
Chickenpox
_______ Native doctors _______ Hospital _______ Other (specify) __________
Fainting
_______ Native doctors _______ Hospital _______ Other (specify) __________
Excessive tiredness
_______ Native doctors _______ Hospital _______ Other (specify) __________
Chest pain
_______ Native doctors _______ Hospital _______ Other (specify) __________
Broken bone
_______ Native doctors _______ Hospital _______ Other (specify) __________
Snake and insect bites
_______ Native doctors _______ Hospital _______ Other (specify) __________
Stomach problems
_______ Native doctors _______ Hospital _______ Other (specify) __________
Madness
_______ Native doctors _______ Hospital _______ Other (specify) __________
Infertility
_______ Native doctors _______ Hospital _______ Other (specify) __________
Sudden sickness
_______ Native doctors _______ Hospital _______ Other (specify) __________
Childbirth
_______ Native doctors _______ Hospital _______ Other (specify) __________
Pregnancy care (antenatal) (If you are a woman where will you recommend?)
_______ Native doctor _______ Hospital _______ Other (specify) __________
Persistent aches
_______ Native doctor _______ Hospital _______ Other (specify) __________
Swollen hand, legs, etc.
_______ Native doctor _______ Hospital _______ Other (specify) __________
Big cut from knife, axe, etc.
_______ Native doctor _______ Hospital _______ Other (specify) __________

Bodily aches and pains
______ Native doctor ______ Hospital ______ Other (specify) __________
Leprosy
______ Native doctor ______ Hospital ______ Other (specify) __________
Deliver a baby (If you are not a woman where will you recommend?)
______ Native doctor ______ Hospital ______ Other (specify) __________

26. How old would you say the head of your immediate family is?
_____ Medium old (40 years or less)
_____ Old (41 to 59 years)
_____ Very old (60 years or more)

27. Are any of your close relatives native doctors?
_____ Yes
_____ No

28. What is the highest level you completed in school?
_____ None (did not go to school)
_____ Primary (elementary) school
_____ Secondary or trade school, or technical school, or commercial school
_____ Higher school, college or technology, A.T.T.C.
_____ University graduate
_____ Completed M.Sc. or Ph.D.

29. For the past two years how many deaths have you had in your own immediate family?
_____ 1 to 2 deaths
_____ 3 to 4 deaths
_____ 5 or more deaths

30. Most of the time do you dress in
_____ Native way of dressing
_____ English way of dressing

31. Are any of your close friends working in a hospital, clinic, or maternity home?
_____ Yes
_____ No

32. Is any member of your immediate family sick at the present time?
_____ Yes
_____ No

33. In general, in a life or death sickness situation where would you go for care?
_____ Native doctor
_____ Hospital or clinic
_____ No place
_____ Other (where?) ______________________________

34. How often have you used native medicine when sick?
_____ Always
_____ Sometimes
_____ Rarely
_____ Never

35. What is your approximate monthly income (salary) after tax?
______ ₦50 or less
______ ₦51 to ₦200
______ ₦201 to ₦400
______ ₦401 to ₦600
______ Over ₦600

36. Have you ever used both the native doctor and the hospital for the same sickness?
______ No
______ Yes
If yes, how often?
______ Once
______ Few times
______ Many times

37. Are you
______ Single
______ Married
______ Divorced
______ Widow/er
______ Separated

38. Would you say that you are between?
______ 21 to 34 years
______ 35 to 49 years
______ 50 to 65 years
______ More than 65 years

39. Is any member of your family suffering from a long illness that does not stop?
______ Yes
______ No

40. If you are married, are you a polygamist? (i.e., have more than one wife)
______ Yes
______ No

41. How much do you like going to the native doctors when you are sick?
______ I like them very much
______ I like them a little
______ I dislike them a little
______ I dislike them very much

42. Are any of your close relatives sick at the present time?
______ Yes
______ No

43. If you have ever been married, how many children do you have?
______ None
______ 1 to 4 children
______ 5 to 12 children
______ More than 12 children

44. Which means of curing sickness do you think is the cheapest?
    ______ Hospital
    ______ Native doctor (medicine man)
    ______ Other (specify) ______________________________

45. When you get sick, what are you most likely to do?
    ______ Immediately look for someone to cure you
    ______ Wait a little and see if you will get better
    ______ Other (specify) ______________________________

46. Does your own immediate family have a car?
    ______ Yes
    ______ No

47. How many children including yourself do your parents have?
    ______ 1 to 4 children
    ______ 5 to 12 children
    ______ More than 12 children

48. Who or which of the following do you live closer to?
    ______ Hospital (clinic) or maternity home
    ______ Native doctor

THANK YOU VERY MUCH

b. Health-Seeking Behavior Survey — Zaire Study

This questionnaire was carefully developed by modifying the Nigerian study instrument to meet the needs of the Zaire study. The Zaire study was funded by the U.S. Agency for International Development (USAID).

HEALTH-SEEKING BEHAVIOR SURVEY INSTRUCTIONS

Please read the following very carefully. Then put a check (√) mark beside the answer that is most applicable for you. Choose only one most correct answer to each question.

Please answer as accurately as possible. There are no right or wrong answers. DO NOT write your name on this paper, because we do not want to know the particular person who answers the questions. Please answer all the questions.

Thank you very much for answering the questions in this form. Your cooperation is highly appreciated. Please ask questions if you do not understand something. Again, thank you for participating.

Immediate family = parents, sisters, brothers, wife/husband, and your children.

Please check (√) only one answer for each question. Indicate who is completing this questionnaire: ______ self
______ interviewer

1. Do you know names of plants that you used to treat malaria?
   ______ Yes
   ______ No

2. If yes, name some of them.

 ____
 ____
 ____
 ____

3. If you have ever used modern medicine when sick, in general, how satisfied were you?
 ______ Very satisfied
 ______ Satisfied
 ______ Dissatisfied
 ______ Very dissatisfied

4. Have you ever used both the native doctor and the hospital for diarrhea?
 ______ No
 ______ Yes

5. If yes, how often?
 ______ Once
 ______ Few times
 ______ Many times

6. Do you know names of plants that you need to treat fever?
 ______ Yes
 ______ No

7. If yes, name some of them.

 ____
 ____
 ____
 ____

8. Do you know names of plants that you used to treat diarrhea?
 ______ Yes
 ______ No

9. If yes, name some of them.

 ____
 ____
 ____
 ____

10. Do you get reimbursed when you buy any modern medicine?
 ______ Yes
 ______ No

11. Are any of your close relatives or a family member ill at the present time?
 ______ No
 ______ Yes

12. How old would you say the head of your immediate family is?
    ______ Medium old (40 years or less)
    ______ Old (41 to 59 years)
    ______ Very old (60 years or more)

13. Have you ever used both the native doctor and the hospital for malaria/fever?
    ______ No
    ______ Yes

14. If yes, how often?
    ______ Once
    ______ Few times
    ______ Many times

15. If you are working for the government, what will you say is your salary level now?
    ______ 0—1,000Z
    ______ 1,000—3,000Z
    ______ 3,000—5,000Z
    ______ 5,000—10,000Z
    ______ over 10,000Z

16. Would you say that you are between
    ______ 21 to 34 years
    ______ 35 to 49 years
    ______ 50 to 65 years
    ______ More than 65 years

17. If not satisfied with traditional treatment received when sick, what do you do?
    ______ Go to the hospital
    ______ Go back to the same native doctor
    ______ Go to another native doctor
    ______ Other (specify) ________________________________

18. What is your religion?
    ______ Catholic
    ______ Protestant
    ______ Muslem
    ______ Kimbanguiste
    ______ None of the above (specify, i.e., write in name) ____________________

19. Where are you living now?
    ______ City or township
    ______ Village

20. Do you know names of modern medicines used to treat malaria?
    ______ Yes
    ______ No

21. If yes, name some of them.
____
____
____
____

22. Which means of transportation do you usually use?
______ Car
______ Motorcycle
______ Tricycle
______ On foot
______ Other (specify) ______________________

23. Which type of water do you usually drink?
______ From the tap
______ From the well
______ Boiled
______ From the river

24. If you have ever used both modern and traditional medicine at the same time for the same illness did you have any reactions such as:
______ Rash or itching
______ Vomiting
______ Death
______ Other (specify) ____________________________

25. Have you a relative, family member, or a friend working in a hospital, clinic, or maternity home?
______ No
______ Yes

26. If you are married, are you a polygamist? (i.e., have more than one wife)
______ Yes
______ No

27. How often have you used native medicine when sick?
______ Always
______ Sometimes
______ Rarely
______ Never

28. What is your tribe?
______ Luba
______ Kongo
______ Mongo/Gbwaba
______ Lunda
______ Tetela
______ Zanda
______ Other (specify) ____________________________

29. When you get sick what are you most likely to do?
______ Immediately look for someone to cure you
______ Wait a little and see if you will get better
______ Other (specify) __________________________

30. How many children including yourself do your parents have?
    ______ 1 to 4 children
    ______ 5 to 12 children
    ______ More than 12 children

31. If you have children would you say that most of your children are:
    ______ Less than 10 years old (born after 1976)
    ______ 10 years old to 20 years old (born between 1966—1976)
    ______ 21 years or more

32. In general, when you use more than 1 (2) medicine for the same illness, which one do you use first?
    ______ Traditional
    ______ Modern

33. Rank the following starting from the most important to the least important thing that helps you to decide where to go for health problems. Use the number 1 as the most important and the number 2 as the second most important.
    ______ Cost
    ______ Seriousness of the problem
    ______ Effectiveness (possibility of cure)
    ______ Distance to be travelled
    ______ Cause of the illness
    ______ Confidence and hospitality
    ______ Other; specify ____________________

34. Most of the time do you dress in
    ______ Native way of dressing
    ______ Western way of dressing

35. Which means of curing sickness do you think is the cheapest?
    ______ Hospital or clinic
    ______ Native doctor (medicine man)
    ______ Other (specify) ____________________

36. If you have already used medicines for the same illness, how did you use them?
    ______ At the same time
    ______ One after the other

37. In general, in a life or death sickness situation where would you go for care?
    ______ Native doctor
    ______ Hospital or clinic
    ______ No place
    ______ Other (where?) ____________________

38. If you have ever used native medicine when sick, in general, how satisfied were you?
    ______ Very satisfied
    ______ Satisfied
    ______ Dissatisfied
    ______ Very dissatisfied

39. If you have used modern medicines against diarrhea, what were the problems, reactions that occurred?

______ Death
______ Rash or itching
______ Vomiting
______ Dizziness
______ Other (specify) ____________________________

40. Why did you use more than one medicine for the same illness?

______ Using more than 1 traditional medicine works better
______ Using more than 1 modern medicine works better
______ Combining modern and traditional medicine works better
______ Someone told me to
______ Other (specify) ____________________________

41. Who or which of the following do you live closer to?

______ Hospital (clinic) or maternity
______ Native doctor

42. If you have ever been married, how many children do you have?

______ None
______ 1 to 4 children
______ 5 to 12 children
______ More than 12 children

43. Whenever you get sick who do you go to first?

______ Native doctor
______ Parents
______ Your husband
______ Someone who works in the hospital
______ The hospital for modern medicine
______ Other (who?) ____

44. Do you know some modern medicines that you used to treat diarrhea?

______ Yes
______ No

45. If yes, name some of them.

____
____
____
____

46. Are you

______ Single
______ Married
______ Divorced
______ Widow/er
______ Separated

47. How often have you used modern medicine when sick?
    ______ Always
    ______ Sometimes
    ______ Rarely
    ______ Never

48. Would you say the head of your immediate family is a
    ______ Man
    ______ Woman

49. What is your occupation?
    ______ Civil Service (not teaching) = Ministry of ______________ (write in)
    ______ Teaching
    ______ Hospital related work (e.g., nursing, hospital lab work, midwifery)
    ______ Housewife (check this only if not working)
    ______ Native doctor
    ______ Farming
    ______ Trading (businessman and woman)
    ______ Unemployed
    ______ None of the above; specify (write in here) ______________

50. Are any of your close relatives, family members, or friends native doctors?
    ______ Yes
    ______ No

51. Are you
    ______ Male
    ______ Female

52. If you have used modern medicines against diarrhea, were you cured?
    ______ Yes
    ______ No

53. Where will you go first for the following?
    Diarrhea or dysentry
    ______ Native doctors ______ Hospital ______ Other (specify) __________
    Malaria
    ______ Native doctors ______ Hospital ______ Other (specify) __________
    Worms
    ______ Native doctors ______ Hospital ______ Other (specify) __________
    Convulsions (fits)
    ______ Native doctors ______ Hospital ______ Other (specify) __________
    High fever
    ______ Native doctors ______ Hospital ______ Other (specify) __________
    Stomach problems
    ______ Native doctors ______ Hospital ______ Other (specify) __________
    Infertility
    ______ Native doctors ______ Hospital ______ Other (specify) __________

54. How close do you live to a hospital or clinic?
    ______ Very close
    ______ Close
    ______ Not close

55. What is the highest level you completed in school?
    ______ None (did not go to school)
    ______ Primary (elementary) school
    ______ Secondary or trade school, or technical school, or commercial school
    ______ Higher school, college or technology
    ______ University graduate
    ______ Completed M.Sc. or Ph.D.
    ______ Other (specify) ________________________________

56. For the past two years how many deaths have you had in your own immediate family due to malaria, diarrhea, or pneumonia?
    ______ 0 (none)
    ______ 1 to 2 deaths
    ______ 3 to 4 deaths
    ______ 5 or more deaths

57. How close do you live to a native doctor?
    ______ Very close
    ______ Close
    ______ Not close

58. Were you cured of malaria when you used modern medicine?
    ______ Yes
    ______ No

59. What do you usually do when you find out that a modern medicine you need is out of stock?
    ______ Come back again and check
    ______ Look elsewhere for the modern medicine
    ______ Use a traditional medicine
    ______ Other (specify) ______________________

THANK YOU VERY MUCH

C. CONTROL AND VALIDITY IN RESEARCH DESIGN

Researchers involved in health services research are often looking at their counterparts in the laboratory, the bench scientists, with some kind of jealousy. This jealousy centers around aspects related to the scope of control that the researcher has over the study. This level of control relates to three relationships. These relationships are the level of control over the strength of the intervention, the level of control over the relationship between the study units comprising the study, and the control of the relationship between the study and the processes going on in the outside world. See Figure 1.

Some examples to clarify the three stipulated relationships and to illustrate their differences within the study design of a laboratory scientist will follow. The first level of control

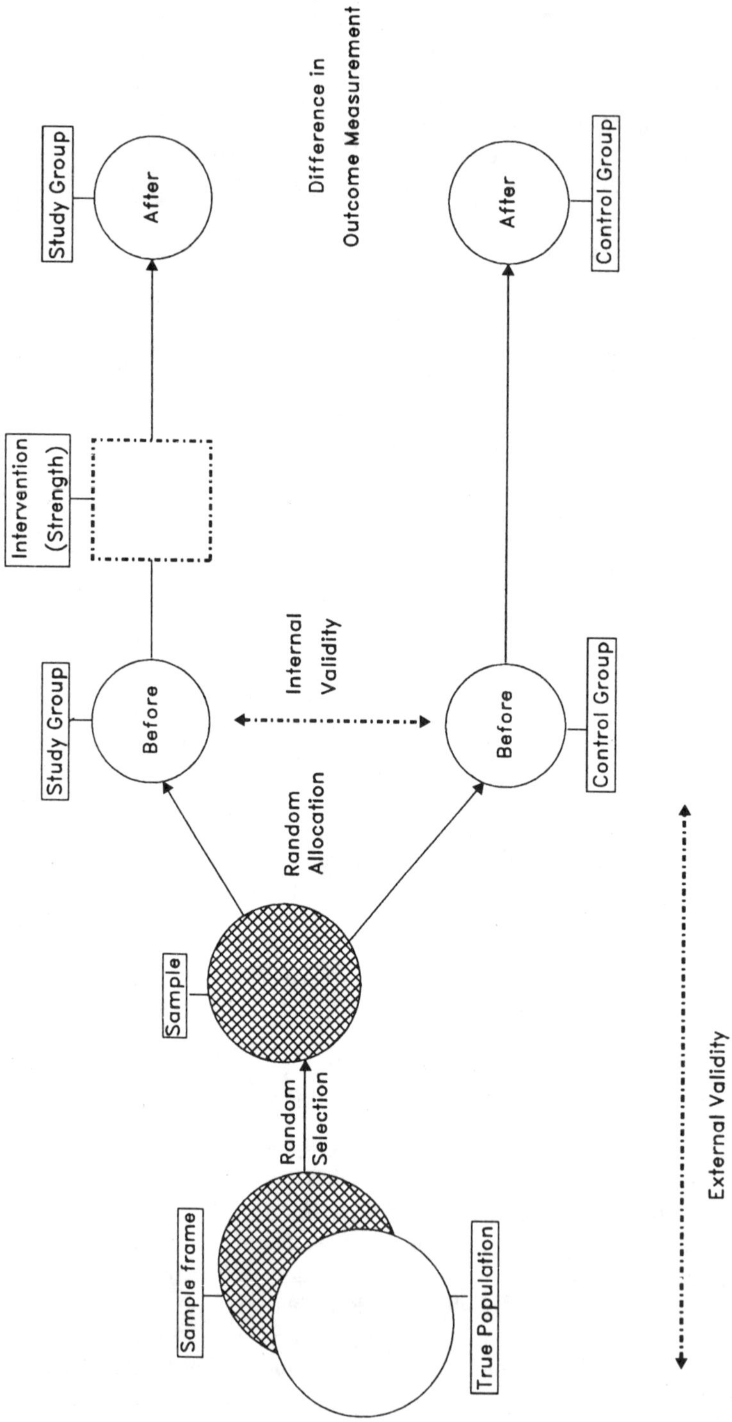

Figure 1. Basic Research Design.

is related to the strength of the intervention. To a large extent, a bench scientist has control over the strength of the intervention in most experiments. For example, the scientist has control over the strength of the solutions used, electric current, and magnetic fields. The strength of the treatment can be measured and controlled easily. To determine the level of control over the delivery of a treatment for a health services researcher is much more complicated. An example of this may be an educational intervention to increase compliance with dental hygienist appointments. The level and intensity of the delivery may vary with the person who provides the educational message, the environment in which the message is delivered (e.g., a noisy office setting), or the way the message is delivered (e.g., a person, a videotape, or written material). The health services researcher may experience, to a larger extent, problems in standardizing the interventions.[5,6]

The second level of control relates to how equal the intervention group is to the control group in the study. To be sure that the intervention works, both control group and experimental group should be the same in composition, and no other factors should exercise any differential influence or any influence at all on either of these groups. In the laboratory we can control the circumstances in which we perform the experiment and take care that these environmental influences are the same. For example, we can control the temperature in which the experiment takes place, the light, the pH, and many other variables. This is not always the case in health services research. Picking up again on our previous example, it may be that while we are doing our compliance study on appointments with dental hygienists a program on television or a newspaper article may point out that gum disease is highly correlated with loss of teeth and the need for dentures. It may be that part of the higher compliance rate found in the study is not due to the intervention, but to the publicity in the newspapers and on television. What is the validity of the findings when we have not controlled for these confounding factors? Control for confounding factors is often referred to in the literature as internal validity.

The third level of control relates to the level of control and expectations we have over the extrapolation of the results of our study to the general population that would have been affected by the study, such as all patients that see dental hygienists. These latter rules are often referred to as external validity. In the following sections, external and internal validity will be discussed in more detail.

1. Control of the Relationship between Study Groups and the Outside World: External Validity

The question arises: what are the elements we have to consider to exercise the highest level of control? The first step in our basic study design calls for maximizing our level of control to make the findings of our study interpretable to the total population. Thus, the study population we select should represent as closely as possible the population of reference. To allow for valid references we should accurately define the sample frame. For example, if my sample frame is the inhabitants of Chapel Hill, NC, would the phone book be an accurate listing of the inhabitants of Chapel Hill? Probably not. If my sample frame definition is all household heads in Chapel Hill, the phone book may be a more accurate sample frame. However, it would not provide us with an optimal sample frame. It may be that two income extremes of our sample frame are underrepresented. Members of the low income group may not be able to afford a phone listing; at the other end of the spectrum, individuals from the upper income group may be underrepresented because the members of this group may ask for an unlisted telephone number. Again, it is essential to define and accurately describe the sample frame from which inferences are made in the study and to obtain an accurate listing of the units of analysis contained in the sample frame.

In the second step to ensure external validity, a representative sample has to be selected from the sample frame. Random sampling methodology as discussed in any standard statistics book can be applied here based on the character of the problem. For two-stage sampling methods, texts that deal with this topic area should be consulted.

The next question to be addressed is: what is the right sample size for the study? The necessary sample size of the study depends on the magnitude of the difference to be measured, the level of significance required, and the power we wish the study to have.[1,4] Often an intervention may be new or the study may be a pilot study, so no precise information is available about the magnitude of the outcome of the evaluation. Then somewhere in the literature the necessary information may be hidden. Someone, somewhere must have done a study with a closely related intervention or, maybe more importantly, by using a different intervention in attempting to achieve a similar outcome, such as increasing scheduled appointment times for the dental hygienist. It is common practice to use the magnitude of the intervention used in such studies to calculate the necessary sample size for our study. Of course, in our study the effect of the planned intervention should be greater or else our intervention would have less value than theirs, controlling for costs.

Having defined our sample frame and made the power calculations to determine the appropriate sample size, our units of analysis have to be assigned to the experimental or the control group. This random assignment is also often called random allocation. Through random allocation of the units of analysis to the control and the experimental group, we attempt to create two groups that are similar, both in the variables that are known to us and which could affect the outcome of the study, as well as in the variables that may affect the outcome of the study but that are unknown to us. An unequal distribution of the known variables could be handled in the final analysis of the study as long as we are able to measure them. However, those unknown covariates, often called confounders, are not known, not measured, and thus cannot be accounted for in the final analysis. It is important to realize that random allocation to control and experimental groups assures a similar distribution of the confounding variables in both the experimental group and the control group.

In addition to the effect that differences in distributions of covariates or confounders have if both groups are different, certain factors or processes in society may differentially affect both of these groups. If our study design can be qualified as quasi-experimental, that means we have selected the control and experimental groups in different ways. The role that these factors and processes play becomes more important.

Some of the common factors which may threaten the external validity are discussed below. The interaction effect of testing, in which a pretest might increase or decrease the subjects' sensitivity or responsiveness to the intervention, alters the perception or nature of the population before we implement the intervention. Normally in society the intervention would be implemented without the pretest and the population would not be sensitized to these issues. Another threat to external validity is the interactive effects of the selection biases and the intervention. We are starting out with two different populations derived from our sample frame. These two populations themselves may be selected subpopulations of the overall sample frame and may differ in characteristics that relate to the study. Another factor that should be considered in studies is the setting (the surroundings, the ambiance). The persons in charge of the intervention may act differently from health personnel in a more realistic setting out in the community. The personnel involved in the administration of the intervention may be more interested in the procedures because of their involvement, and personnel may be better trained. The environment in which the intervention is implemented may be different from the actual practice setting.

Another threat to external validity is multiple treatment interference. This may happen when the effect of one treatment is not washed out before the next intervention is implemented or when one intervention prepares the subject for the following intervention. Crossover designs may offer such a problem. In crossover designs the intervention group serves as its own control. External validity in research designs can be assured by employing the proper sampling techniques and procedures.

2. Control over the Relationship between the Control Group and the Experimental Group: Internal Validity

Internal validity is threatened by approximately eight different factors. We will follow the nomenclature used by Campbell and Stanley[2] and Cook and Campbell[3] in discussing these factors.

The first factor is named *history*. History refers to all activities and forces that are going on in society at large, or in the particular study setting, between the pre- and posttest measurement and that may alter the outcome of the study. If our control group and experimental group are the same on the covariates under consideration, then history will have little effect. If, however, both groups are different in composition, the history may influence both groups differentially and change the outcome of the study. For example, if the dental association in our appointment compliance study was to launch a large public campaign urging the public to see the dentist between the "before" and "after" observations, this might alter the outcome of our study. Certainly when an interactive process with the intervention can be observed, such as a reminder postcard, those individuals that are receiving the reminder are more likely to follow up on their visit in the preventive dental care program.

The next factor is called *maturation*. This refers to processes that are going on in both control and experimental populations and are often related to the study's duration. The longer the study takes to complete, the more important it becomes to guard for threats of maturation. Maturation refers to processes that go on in units of analysis because of elapsing time. An example in our case would be that a population which is growing older and becoming more prone to gum disease would be more likely to comply with scheduled visits to the dental hygienist.

Testing is another process that can influence the outcome of the study. People learn how to respond to a test. When the same test is being used in the premeasurement observations as well as in the postmeasurement observations, then individuals who are blinded to the study may be able to "out-psych" the outcome of the study. As everyone is aware, most tests reveal their purposes to the attentive reader. In addition, people learn to score on a test. Individuals enrolled in studies may want to please the investigators and will fill out the test instrument the way they think the researcher would like to have it completed, introducing a fallacy in the outcome of the study.

Instrumentation is another important factor to consider in the study design. Changing instruments throughout the study may introduce fallacies in the outcomes of the study. For example, if one is evaluating the blood pressure lowering effects of a drug and uses different gauges for control and experimental groups or before and after the intervention, these instruments may not all be calibrated in the same manner or show the same results in comparisons. Instrumentation considerations must be built into the overall study design.

The problem of *statistical regression* is often encountered because we researchers like to improve or change those elements that are the worst. We like to achieve improvements, sometimes out of compassion, or think we will achieve the greatest impact where the need for our intervention is the highest. However, there are some problems connected with this approach. The concept came into use from the area of educational research. Many programs in educational research are based on improving the performance of low performers in the classroom. Low performers are often selected based on testing results, but as anyone who has ever taken an SAT test knows, there is often a variability in the outcome of the testing. Sometimes one has a good day and does relatively well, and sometimes one has a bad day and performs badly. Among our group of bad performers, we may have a group of individuals who are actually performing much better than the others and will perform much better on the next test than they did on the first test. Thus, chances are we will see an improvement in the performance of our students that is not due to the intervention, but may in some way be due to our selection process. A similar concern may be expressed for selecting hypertensives or other study populations based on a one-time measurement.

Biases result in the selection of different populations for control and experimental groups. In a randomized trial these biases will not occur, but if control and experimental groups are composed based on criteria different from randomization, biases can be introduced by the investigators in selecting the individuals for the control and experimental groups.

Experimental mortality refers to the drop-out rate of the individuals enrolled in the study. The drop-out rate should not only be even in both groups, but the drop-outs in both groups should be equal in all variables that may have an effect on the outcome of the study. Again, it may be that the drop-out rate is not evenly distributed, but there is a higher drop-out rate in the experimental group than in the control group. It can often be assumed that the drop-out rate is higher in the experimental group due to the intervention. The intervention may be painful or require some activities or put other demands on the individual. In the final analysis the investigator should describe the drop-out rate and provide reasons why the individuals in the drop-out group are not different from the remaining subjects in the experimental group as well as the control group.

Furthermore, most of the factors that may affect the interpretation of the results described above may provide some form of interaction. For example, we may have selected different individuals for the control group as well as the experimental group. A factor such as maturation may take place at a different rate in the control group vs. the experimental group due to differences in the composition of these populations. The interaction may differentially affect both study populations. Careful consideration has to be given in any study design to factors, often called confounders, which may alter the outcome of the study.

REFERENCES — SECTION II.C

1. **Fleiss, J. L.,** *Statistical Methods for Rates and Proportions,* 2nd ed., John Wiley & Sons, New York, 1980.
2. **Campbell, D. T. and Stanley, J. C.,** *Experimental and Quasi-Experimental Designs for Research,* Rand McNally, Chicago, 1963.
3. **Cook, T. D., and Campbell, D. T.,** *Quasi-Experimentation, Design and Analysis Issues for Field Settings,* Rand McNally, Chicago, 1979.
4. **Cohen, J.,** *Statistical Power Analysis for the Behavioral Sciences,* Academic Press, New York, 1969.
5. **Sechrest, L., West, S. G., Phillips, M. A., Redner, R., and Yeaton, W.,** Some neglected problems in evaluation research: strength and integrity of treatments, in *Evaluation Studies Review Annual,* Vol. 4, Sage Publications, Beverly Hills, CA, 1979, 15.
6. **Yeaton, W. H. and Sechrest, L.,** Critical dimensions in the choice and maintenance of successful treatments: strength, integrity and effectiveness, *J. Consulting Clin. Psychol.,* 49, 156, 1981.

III. SAMPLING METHODS AND SAMPLE SIZE

The selection of a sample in a population is an important phase in any research that utilizes sampling. A sample is a small representative proportion of a population that is selected randomly for analysis. Random selection is very important in sampling since it is upon this that most of the principles of statistical inferences are based.[1,7]

A. TYPES OF SAMPLING

1. Probability Sampling

In probability sampling, the chance that an element in a target population will be selected is known. As a result, the sample is representative of the population. Probability sampling includes simple random sampling, systematic sampling, stratified random sampling, and area or cluster sampling. Table 2 compares these different probability sampling designs.

TABLE 2
Comparisons of Probability Sampling Designs

Type of design	Population characteristics	Example of population type	Advantages	Disadvantages
Simple random	Generally homogeneous or homogeneously conglomerate	Hispanic expectant mothers to examine their level of AIDS knowledge	Process is simple and representative of population	May not be feasible to accurately identify entire population
Systematic	Ordered elements according to predetermined sequence	50 students out of 500 to determine attitude toward homosexuality	Representative and less time-consuming than simple and stratified	Process might skip a significant portion of the population that are grouped together on the sequence
Stratified	Differences between definite strate out homogeneity within each stratum	A city whose population is African Americans, Hispanics, Asians, Native Americans, and Whites	Researcher can control for population differences that may influence results of study	May not be feasible to identify the number of elements in each stratum precisely
Area or cluster	Clusters are similar in their characteristics yet heterogeneous within each unit	Selection of 300 teachers from a nation's schools (all schools similar in their curriculum, mission etc. yet students and teachers who are in the schools are different in age, gender, philosophy, etc.)	Allows random sampling of a large population that is impossible with simple random sampling	Multiple stages of sampling increases sampling error; data from clusters may not be representative of population

a. Simple Random Sampling

Sampling is categorized as simple random when every element in the population has an equal probability of being selected. Simple random sampling is the foundation on which probability sampling is built.[8] Each selection in a simple random sample is independent of other selections. The procedure consists of identifying a population which is characteristically homogeneous or homogeneously conglomerate.

Simple random sampling is sampling with or without replacement. Depending on the sampling technique, the probability for the selection of each element in a population is different. Selection with replacement is the most conservative random sampling method. It provides an equal chance for each element to be selected. For example, if the population comprises 500 people, each person has a 1 in 500 chance of being selected. In this method, if the researcher draws names out of a hat to obtain 50 subjects, each name must be replaced before the next name is drawn to ensure an exactly equal chance for each subject. In sampling without replacement, each element has a different level of probability for selection. For example, if the researcher is selecting 50 subjects from a population of 500, the first name has a one in ten chance, or a 0.1 probability, of being selected. If the first name is not replaced, the second name has a 49 in 499 chance, or a 0.098 probability, of being selected. The probability for selection decreases as further names are drawn.

There are several methods for achieving random selection: for example, computer, roulette wheel, or the flip of a coin. However, the most common method of random selection is the use of a table of random numbers.* To identify a starting position on the table, the

* Please refer to any statistics textbook for an example of a table of random numbers.

researcher places a pencil, pen, or finger on the random table while the eyes are closed. The researcher can then move up, down, right, or left to select numbers until the desired sample size is obtained.

b. Systematic Sampling

Another type of random sampling is systematic sampling. This process involves the selection of certain elements in a series according to a predetermined sequence. A precondition for conducting this process is the availability of an ordered list of all elements in the sample frame, i.e., members of a population. The researcher must know the number of people in the population and the sample size desired. Assuming the researcher wants a sample of 50 from a population of 500, the population size is divided by the desired sample size, yielding "K", which is the gap between persons selected from the ordered list. In this case, N = 500 divided by N = 50 is K = 10. Systematic sampling in this case is the process of selecting every 10th person on the list until a total of 50 are selected. A precondition for this process to be accurate is that the ordered list must be random. One drawback in systematic sampling is that the process might skip a significant portion of the population that is grouped together on the list. Thus, the accuracy of this process is more dependent upon the adequacy of the sampling frame than in simple random sampling. When sampling bias is evident the researcher must randomize the sampling frame or use simple random sampling.

c. Stratified Random Sampling

In the stratified random sampling method, the population is subdivided into smaller homogeneous groups so as to attain a more accurate representation. The population is divided into strata based on known variables before the sample is selected at random.

Variables used to stratify populations are mostly demographics such as age group, race, ethnicity, nationality, gender, socioeconomic status, geographic region, type of institution, grade level in school, type of care, and type of program.

Stratified sampling may result in either proportionate (when the number of elements is selected in proportion to their occurrence in the population) or disproportionate sampling (when each stratum has an equivalent number of elements in each sample regardless of their occurrence in the population). For example, in a sample size of 50, suppose that the population has a racial distribution of 300 whites, 100 blacks, 80 Hispanics, and 20 Native Americans. A proportionate stratified sampling would require that 10 people be selected randomly from each racial group for a total sample of 50. In contrast, a disproportionate stratified sampling would require that the random selection of each stratum be based on a certain percentage of the total proportional population. In this case, since N = 50 is 10% of N = 500, 10% of 300 = 30 for white; 10% of 100 = 10 for blacks; 10% of 80 = 8 for Hispanics; and 10% of 20 = 2 for Native Americans.

In the above example the study would only include two Native Americans, thus resulting in underrepresentation. On the other hand, the equal number of subjects (ten) means that the white group is also underrepresented. In disproportionate stratified sample design, weighing factors are used to prevent inequalities in selection probabilities from causing the introduction of sample bias.

For stratified random sampling to be accurate, the researcher must know the variables in the population that are critical to attaining representativeness.

d. Area or Cluster Sampling

The cluster or area sample is particularly appropriate when a simple random sample proves to be prohibitive in terms of travel time and cost. It is useful when (1) the target population is infinite, (2) the geographic distribution of the individuals is widely scattered, and (3) a list of the members of the population does not exist.

When the individual elements that make up the sampling frame are not known it is difficult to develop a sampling frame. Random selection is only possible by multistage sampling. In cluster sampling a sampling frame may begin with a list of all states, counties, cities, institutions, or organizations with which elements of the identified population would be linked.

In a multistage cluster sampling in Nigeria, from the 21 states in the country, the investigator can randomly select a sample of 7. From the 7 states, all local government areas (equivalent of counties in the U.S.) could be listed, and a random sample of 50 local government areas (LGAs) could be selected. From the 50 LGAs, all of the secondary schools (7 through 12 grades) could be listed and a random sample of 25 schools selected. A list of all the teachers in the 25 schools could be obtained and a random sample of 300 teachers selected. This multistage random sampling provides the opportunity for obtaining a larger sample from a diverse population at a lower cost.

While cluster sampling has some advantages over simple random sampling, there are some disadvantages. Since there are two or more stages of sampling, there are more chances for sampling error. Additionally, there may be sample bias because of the unequal size of some of the subject groups selected. The use of intact groups increases the chances of having a sample that is not typical of the entire study population; thus, data from elements associated with the same subject are likely to be correlated rather than completely independent.

2. Nonprobability Sampling

In this sampling method the chance that an element in a target population will be selected in the sample is unknown. As a result, the sample is not representative of the population. Research results obtained in nonprobability samples are treated as suggesting statistical characteristics about a population rather than as generating population estimates with a specifiable level of confidence.[5] Since the process is nonrandom, the ability to generalize results beyond the sample is greatly limited. However, the sampling method is less expensive and simpler to apply; hence, it is valuable for small investigations and pilot studies, particularly when used to improve validity and reliability of questionnaires. There are several types of nonprobability samples. They include accidental or convenience sampling, judgment or purposive sampling, network or snowball sampling, quota sampling, dimensional sampling, and mixed sampling designs.

a. Accidental or Convenience Sampling

In accidental or convenience sampling, the approach is based on selecting elements in a population that happen to be at the right place at the right time. Elements are selected from the target population on the basis of their accessibility or convenience. A common example of convenience sampling is selecting subjects from a classroom until the desired sample is attained. The major assumption associated with this type of sampling is that the elements in the target population do not differ based on accessibility or convenience. Sampling biases tend to be present in this type of sampling. Subjects who choose to participate may have characteristics different from others. The setting for the study may exclude minorities or the elderly. Students in a health education class have different levels of knowledge, attitudes, and behavior than students in non-health-related classes. Smokers who choose to enter a smoking cessation program are different from those who choose not to enter a program. Accessible elements within a target population may differ significantly from those less accessible elements.

However, a researcher can minimize sampling bias in this method by identifying and describing known biases in the sample. Accidental or convenience sampling is inexpensive, less time-consuming, and accessible. Their limitations notwithstanding, they provide means for obtaining information in topics or areas that are difficult or sometimes impossible to explore using probability sampling.

b. Judgment or Purposive Sampling

Judgment or purposive sampling involves the selection of elements which represent a "typical" sample from a target population. When this approach is used, the quality of samples selected depends on the accuracy of the researcher's judgment of what constitutes a typical sample. This approach has been criticized because in the absence of an external criterion it is impossible to evaluate the precision of the researcher's judgment. However, this method is useful in qualitative studies when a researchable hypothesis needs to be explored where other sampling techniques prove difficult to apply.

c. Network or Snowball Sampling

Network or snowball sampling involves a multistage technique that utilizes social networks of individuals who tend to share common characteristics. The researcher must first identify and interview a few subjects with requisite criteria. These subjects are then asked for their assistance in identifying others with the requisite criteria. These persons are then interviewed and maybe be asked to identify others until a satisfactory sample is obtained. This procedure is very useful for finding subjects who may not be willing to make themselves known in the population. Examples include homosexuals, alcoholics, drug addicts, sex offenders, and child abusers. As with other nonprobability sampling procedures, there are obvious biases, since the subjects are not independent of each other.

d. Quota Sampling

Quota sampling procedure involves the selection of proportional samples of subgroups within a target population to ensure generalization of findings. This method is the nonprobability equivalent of a stratified random sample. Quota sampling ensures the inclusion of population subgroups that are likely to be underrepresented. Examples include racial minorities, the elderly, the poor, and the very rich. This sampling method provides more accurate sampling compared to accidental or convenience sampling because mathematical weighting can be used to minimize sampling biases. Although restrictions are placed on the number of subjects selected from each subgroup, there is often little or no control exercised over the selection of subjects within each subgroup. Researchers can minimize these biases by sampling for representativeness within each subgroup rather than for accessibility and convenience.

e. Dimensional Sampling

In dimensional sampling the population is delineated on the basis of requisite characteristics. This approach is primarily a multidimensional form of quota sampling. At least one case is selected to represent each of the characteristics. In this sampling technique only a small sample is needed, since each selected case will be examined in more detail.

f. Mixed Sampling Designs

Mixed sampling designs constitute the combination of both probability sampling and nonprobability sampling procedures. Nonprobability methods (such as judgmental sampling; see Table 3) are usually employed in the early stages, while probability sampling procedures are employed in the latter.[2] These sampling designs are used particularly in research that utilizes the combined qualitative and quantitative methods (triangulation). This has been very useful when focus group interviews are conducted to complement surveys in research.

B. SAMPLE SIZE

One of the major concerns of a beginning researcher is determining the appropriate sample size to use in a study. Sample size depends upon the extent to which the sample population represents the requisite criteria of certain qualities and characteristics in the general population. Although there are no absolute qualifiers to be used in judging optimum sample

TABLE 3
Comparisons of Nonprobability Sampling Designs

Type of design	Population characteristics	Example of population type	Advantages	Disadvantages
Accidental or convenience	Accessible and convenient population	Classroom students	Inexpensive, accessible, and less time-consuming	Accessible elements may differ significantly from those that are less accessible
Judgment or purposive	Population that represents a "typical" sample	A community in which the populaiton has characteristics markedly different from others	Useful in qualitative studies	Difficult to evaluate the precision of a researcher's judgment
Network or snowball	Social networks of individuals that share common characteristics	Groups that are not willing to make themselves known, e.g., alcoholics, homosexuals, and child abusers	Utilizes process that allows access into populations that would otherwise be inaccessible	Bias in the selection process; those willing to participate may be significantly different from others
Quota	Differences between subgroups, but homogeneity within each group	Subgroups that are likely to be underrepresented, e.g., elderly, ethnic minorities	Allows broader representation of every subgroup in a population	Little or no control in the selection of subjects within each group
Dimensional	Multidimensional form of quota sampling	Same as quota sampling	Same as quota sampling	Since small sample is usually involved, generalization is difficult to make
Mixed	Large population where only partial characteristics are known	Heads of households in a rural areas in Pennsylvania	Opportunity for use of multiple research designs	Process can be cumbersome

size, there are formal procedures used to determine the adequacy of the sample size in a given study. In other words, the determination of adequate sample size must be specific to the study. One general rule is to make the sample size as large as possible to ensure a greater likelihood of adequate representation. Regarding the characteristics of the population, a larger sample size will be needed in a markedly heterogeneous population than in a homogeneous one.[9,10] Additionally, constraints such as time, money, and availability of subjects are important considerations in the decision about sample size. The researcher should consider these eight important factors in making decisions about sample size:

1. Type of study. Descriptive and correlational studies often use large samples. Experimental and quasi-experimental studies often use smaller samples than descriptive and correlational studies. Exploratory studies, whether qualitative or quantitative, and case studies tend to use very small samples.
2. The variability of the population. This is usually expressed as the standard deviation. As variability increases the needed sample size increases.
3. The number of variables. As the number of variables under investigation increases the required sample size increases.
4. The level of precision. This is the degree to which the sample population will adequately represent the general population.

5. The sensitivity of the measurement tools. This is the strength of the validity and reliability of the instrument. In cases where scores tend to vary from the actual situation because of a weak instrument, a larger sample size is needed to enhance an understanding of the phenomenon that is being examined.
6. The sampling method employed. Using a probability or nonprobability sample and the selected sampling procedure will influence the decision on sample size.
7. Expected effect size. Expected effect size refers to the expected difference in scores in an experimental treatment group. This difference may be large or small but meaningful. A large effect (associated with a large sample) is much easier to detect statistically than a small effect (associated with a small sample).
8. Data analysis techniques. This refers to the ''power'' (the ability to detect differences) of the statistical analysis. The researcher should use a large sample when the power of the planned statistical analysis is weak. Although parametric statistics are more powerful in analyzing differences than nonparametric statistics, nonparametric statistics are more powerful when a small sample has been used.

1. Determining Sample Size

An acceptable representativeness of the sample based on an established confidence level and sampling error can be calculated using the following formula:

$$N = \left(\frac{Z}{e}\right)^2 (p)\,(1 - p)$$

where N = sample size, Z = the standard score corresponding to a given confidence level, E = the proportion of a sampling error in a given situation, and P = the estimated proportion or incidence of cases in the population. When calculating the sample size the probability of error in using a sample instead of the entire population must be considered. In other words, how far would the sample mean deviate from the population mean? This is referred to as the standard error of the mean. This usually is determined statistically as follows:

$$Se_{\bar{x}} = \frac{S}{N}$$

where $Se_{\bar{x}}$ = the standard error of the mean, S = the standard deviation of the sample, and N = the number of units in the sample. This statistical procedure for computing the standard error of the mean can be used for both small and large samples.

2. Sampling in Minority Populations

Sampling design for a specific minority population is the same as in a nonminority population. However, when a study is designed to compare and contrast a minority population with a nonminority population or to look at a population of minorities and nonminority, the simple random technique will be inappropriate due to the disproportionate representation of the different groups. Consequently, a systematic sample or cluster sampling is used.

Assume that a sample of 400 participants in a study (200 blacks and 200 whites) were being selected for a study in Philadelphia, PA.

According to the 1980 National Census Distribution Report, of a total of 367 census tracts in Philadelphia, 2 tracts had a 100% density of black households, 32 tracts had a 95% density, 23 tracts had a 90% density, etc., to a total of 182 tracts that had no black residents. The density within census tracts suggests the percentage of blacks that are to be sampled. For example, 90% density indicates that, of 100 samples, 90 will be blacks by probability. The limitation of selecting high-density tracts may be that we select only inner-city highly clustered blacks. However, if we set a low density we have a lower efficiency of reaching blacks. To make the selection procedure practical and workable, 60 to 100% density tracts are designated to draw the black samples. By this probability, approximately 85% of blacks will be drawn out of every 100 samples. Accordingly, drawing 240 samples, approximately 200 blacks will be reached. The other 40 may be whites who will be included in the 200 white sample. The other 160 whites as required can be drawn randomly from 231 census tracts which have ≤40% density blacks. By probability, approximately 85 white households will be drawn out of every 100 samples. By sampling 200 samples, 160 whites should be selected. Blacks drawing from these 200 samples can be eliminated.

Three major factors will be taken into account to determine the total number of samples in order to complete 200 blacks and 200 whites in Philadelphia:

1. Refusal rate. Frey,[3] in reviewing 182 survey interview studies, has suggested that a median refusal rate of 28% might be expected.
2. Nonincidence rate. Nonincidence rate refers to the household(s) that may have a language barrier, a hearing problem, or some other reason that would render them unable to complete the interview. This rate has been estimated to be 10%.
3. No-answer household rate. This means that the respondent may be out of town or otherwise unavailable. This rate has been estimated to be 25%.

Considering these three factors, it is estimated that each sampled household should have an average of two substitute (replicate) households. Thus, 720 black households and 600 white households will be sampled. Since all samples are generated randomly, it is not difficult to assign substitute households.

3. Selection of Multiple Respondents in a Household

If a household contains more than one eligible respondent, it can be regarded as a cluster of sampling units, and an additional selection procedure must be implemented. This is particularly true for the present general population survey, where the only qualification for participation is to be an adult over the age of 18. Troldahl and Carter[12] developed a selection procedure which is easy to use. The interviewer begins by asking two questions to the household head:

1. Including yourself, how many persons 18 years or older live in your household?
2. How many of them are men?

The interviewer then follows an explicit procedure for selecting the respondent, wherein four categories are randomly assigned to household numbers (see Appendix).

C. MEASUREMENTS

Measurement is the concept of quantifying and evaluating phenomena based on preestablished standards. To attain meaningful results in research, the design and implementation must be based on measurable phenomena.[4,13] Two critical concepts of measurement are vital in research design; they are validity and reliability.

1. Validity

Validity as a concept of measurement refers to the degree to which a research procedure or tool measures what it is supposed to measure. In establishing validity measures there are six basic approaches used to obtain an accurate measure of the phenomenon under study. They are face validity, content validity, construct validity, concurrent validity, predictive validity, and criterion validity.

a. Face Validity

Face validity refers to the degree to which an item in a measuring instrument "appears" to measure what it is supposed to measure. When developing or adopting a questionnaire, one must be careful and ensure that the items are specific enough to assess the desired variable.

b. Content validity

Content validity refers to the degree to which the content of an instrument proportionately matches the preestablished objective criteria of a concept. This approach is used to ensure that an evaluation of a phenomenon is in accord with preestablished objective criteria. For example, if the syllabus of an emergency care course requires that the study spend one third of the time on skills, at the end of the course one third of the evaluation (test) should cover skills. If only one fourth of the test is on skills, it will be low on content validity.

c. Construct Validity

Construct validity has to do with how one measure relates to other measures. It refers to the extent to which measures of a phenomenon have a pattern of correlations with other measures as an expected response or outcome. Construct validity is high when two things are established: (1) when a score on a scale is correlated with other measures with which it is predicted to correlate (convergent validity) and (2) when a score on a scale fails to correlate with measures from which it is supposed to be different (discriminant validity). For example, people with suicidal tendencies are expected to experience varied degrees of depression.

d. Concurrent Validity

Concurrent validity refers to the correlation between scores of two different scales or measures established at about the same time. For example, a student who has high SAT scores and a high grade point average at about the same time will establish a high concurrent validity. Students who score high on nutritional knowledge and a practice test and about the same time, upon unobstrusive observation, are able to select food in the cafeteria from the four basic food groups will establish a high concurrent validity.

e. Predictive Validity

Predictive validity means the extent to which scores on a scale or test predict future behavior. This can be measured by a method of correlation coefficients between individual scores on a scale or test and later behavior. If students have high scores on a nutritional knowledge, attitude, and practice test and are later found to be uninterested and unable to select food in the cafeteria from the four basic food groups, the nutrition scale is said to have low predictive validity. The difference between concurrent validity and predictive validity is that in the former the new scale is correlated based on a criterion measure that is given at about the same time, while in the latter the criterion measure is a performance or behavior that occurs in the future.

f. Criterion Validity

Criterion validity can be either a concurrent criterion validity or a predictive criterion validity. Criterion validity is the process of comparing the cost effectiveness of the application of a costly instrument against a less costly one. In the correlation between the two instruments the criterion validity of the less costly instrument is considered high. Concurrent criterion validity refers to the use of the two instruments at the same time, while predictive criterion validity refers to using the less costly instrument at one time to predict the measures of the more costly instrument at the second time.

2. Reliability

This refers to the degree to which an instrument accurately and consistently measures a given phenomenon from one time to another. An instrument with high reliability will yield comparable scores upon repeated administration. A high coefficient of reliability means a low error of measurement.

An instrument of measurement may be reliable even though it is not valid, but a valid instrument is always reliable.

a. Methods of Determining Reliability

There are several ways of computing the reliability of a measuring instrument. For some instruments the reliability has been established by the original designer. However, most instruments have no established reliability.[11] Some of the common methods of assessing reliability include internal consistency, multiple form, test/retest, interobserver, and intraobserver.

Internal Consistency — This method measures the reliability of a self-reporting instrument (questionnaire, test) that is administered at the same time. Internal consistency measures include:

1. KR-20 — reliability measure for instruments which have dichotomous items, such as yes/no and right/wrong.
2. Split half — this approach measures reliability by randomly assigning all items in an instrument to two groups, such as odd and even. The sums of the scores of the two groups are correlated since both groups are measuring the same construct; a reliability level of 0.80 or higher is expected.
3. Cronbach's α-coefficient — this statistical measure approximates the mean of all possible split-half correlations.

Multiple Form — In this approach, subjects respond to two instruments (or two forms of the same instrument) that measure the same construct at the same session. Again, as in split half, a correlation of 0.80 or higher is expected for high reliability since the instruments are measuring the same construct.

Test/Retest — This method measures the reliability of a self-reporting instrument that is administered at different times. The reliability scores in this measure are usually lower than split half or multiple form because of the time lag between the two different measuring times. The longer the time span between the measuring times, the greater the likelihood of a low reliability.

Interobserver — This approach involves the observation of a phenomenon or performance by more than one person at the same time. If there are three observers involved, their performance can be rated and correlated.

Intraobserver — This approach involves the observation of a phenomenon or performance at different times. If behavior is observed at two different measuring times, a comparison of the two observations can be made to test consistency and, thus, reliability. This measure is sensitive to changes in time. Because of concern over the impact of time change on an observed outcome, the interobserver approach is preferable to the intraobserver.

3. Levels of Measurement

Data are subjected to one form of measurement or another. The characteristics of measurability are classified into four basic measurement scales. They are the nominal scale, the ordinal scale, the interval scale, and the ratio scale.

a. Nominal Scale

In the nominal scale, measurement is obtained by classifying specific groups of objects or elements together and assigning a name to them. For example, a classroom of students can be classified into two groups — male and female. The nominal scale is a crude measure used to quantify elements into groups. Numbers in the nominal scale have no relationship to one another and are useful in the grossest measure of central tendency — the mode. The bar graph is an example of how a nominal scale may be reported graphically.

b. Ordinal Scale

In the ordinal scale, measurement is obtained through some ordered ranking of groups in categories. However, the distance between each ranking is not known. A student who ranks second in the class is definitely higher than a student who ranks seventh. However, the distance between these two students is not known. The ordinal scale can be used in two measures of central tendency — median and mode. The data can be reported graphically as a line graph, a percentile scale, and a histogram. Such variables as weight, height, and family income are ordinal data.

Both nominal data and ordinal data are categorical data and use nonparametric statistics.

c. Interval Scale

The interval scale deals with the specific amount of difference between scores. In this scale there is no absolute zero point. Data are measured in terms of more or less and how much much or how much less. For example, 3.6 is 0.1 less than 3.7. Interval data are generated from many attitude and aptitude scales. The thermometer is an example of an interval scale; 30° is colder than 31°. Using a depression inventory, although one may not be twice as depressed, the fineness of the graduation between the equidistance points in an attitude (Likert-type) scale might improve measurement.

d. Ratio Scale

The ratio scale is an interval scale with a true zero point. A ratio scale involves the measurement of one factor in relation to another. The celsius thermometer scale is an example of a ratio scale because it begins with a zero.

Ratio data are usually reported graphically on a grid. One axis measures the strength of one factor while the other axis measures a second factor. The two factors intersect at a point which represents the relationship of both phenomena to each other.

REFERENCES — SECTION III

1. **Best, J. W.,** *Research in Education,* 3rd ed., Prentice Hall, Englewood Cliffs, NJ, 1977.
2. **Burns, N. and Grove, S.,** *The Practice of Nursing Research Conduct, Critique and Utilization,* W. B. Saunders, Philadelphia, 1987.
3. **Frey, J. H.,** *Survey Research by Telephone,* Sage Publishing, Beverly Hills, CA, 1983.
4. **Green, L. W. and Lewis, F. M.,** *Measurement and Evaluation in Health Education and Health Promotion,* Mayfield Publishing, Palo Alto, CA, 1988.
5. **Hess, I.,** *Sampling for Social Research Surveys, 1947—1980,* Survey Research Center Institute for Social Research, University of Michigan, Ann Arbor, 1985.
6. **Leedy, P. D.,** *Practical Research Planning and Design,* 2nd ed., Macmillan, New York, 1980.
7. **Ross, K. N.,** Sample design, *Int. J. Educ. Res.,* 11(1), 57, 1987.
8. **Rubinson, L. and Neutens, J. J.,** *Research Techniques for the Health Sciences,* Macmillan, New York, 1987.
9. **Sconun, M. J.,** *Sampling. A Quick, Reliable Guide to Practical Statistics,* Simon and Shuster, New York, 1960.
10. **Slavin, R. E.,** *Research Methods in Education. A Practical Guide,* Prentice Hall, Inglewood Cliffs, NJ, 1984, 75.
11. **Sowell, E. and Casey, R.,** *Analyzing Educational Research,* Wadsworth Publishing, Belmont, CA, 1982.
12. **Troldahl, V. C. and Carter, R. C.,** Random selection of respondents with households in phone surveys, *J. Marketing Res.,* 1, 71, 1964.
13. **Windsor, R. A., Baranowski, T., Clark, N., and Cutter, G.,** *Evaluation of Health Promotion and Education Programs,* Mayfield Publishing, Palo Alto, CA, 1984.

APPENDIX

Q-1 The first one is, how many people 18 years and older live in this household, including yourself?__________________

CIRCLE ANSWER IN THIS ROW

Q-2 How many of them are men?

		1	2	3	4
	0	Woman	Oldest woman	Youngest woman	Youngest woman
CIRCLE	1	Man	Man	Man	Oldest woman
ANSWER IN	2		Oldest man	Youngest man	Youngest man
THIS	3			Youngest man	Oldest man
COLUMN					
					Oldest man

(INTERVIEWER CIRCLE CATEGORY AT INTERSECTION AND USE IN THIS SENTENCE)
Okay, according to the method used by our university, I need to interview the __________________ in your household. (age-sex)

Categories Used for Selecting Respondents

Total number of men in household	Total number of adults in household			
	1	2	3	4
Version 1				
0	Woman	Oldest woman	Youngest woman	Youngest woman
1	Man	Man	Man	Oldest woman
2		Oldest man	Youngest man	Youngest man
3			Youngest man	Oldest man
4+				Oldest man
Version II				
0	Woman	Youngest woman	Youngest woman	Oldest woman
1	Man	Man	Oldest woman	Man
2		Oldest man	Woman	Oldest woman
3			Youngest man	Woman or oldest woman
4+				Oldest man
Version III				
0	Woman	Youngest woman	Oldest woman	Oldest woman
1	Man	Woman	Man	Youngest woman
2		Youngest man	Oldest man	Oldest man
3			Oldest man	Youngest man
4+				Youngest man
Version IV				
0	Woman	Oldest woman	Oldest woman	Youngest woman
1	Man	Woman	Youngest woman	Man
2		Youngest man	Woman	Youngest woman
3			Oldest man	Woman or youngest woman
4+				Youngest man

From Troldahl, V. C. and Carter, R. C., Random selection of respondents within households in phone surveys, *J. Marketing Res.*, 1, 71, 1964. With permission.

Chapter 4

DATA ANALYSIS AND SELECTION OF STATISTICAL METHODS

M. Wang, C. O. Airhihenbuwa, and E. Nnadi Okolo

TABLE OF CONTENTS

* E. Nnadi Okolo.
** M. Wang.
***C. Airhihenbuwa.

I. INTRODUCTION

Data analysis is the cornerstone in reporting research findings. The importance of selecting appropriate statistical methods in analyzing research data cannot be overemphasized. The interpretation of research data is only as good as the statistical analysis. The results of the best-planned and organized, most expensive, well-thought-out research study will be less useful, if not useless, if the wrong statistical methods are used in analysis. The nature

* E. Nnadi Okolo.

of data obtained from a research study will determine the type of statistical analysis to be used. When a research study is exploratory (often used in developing a theory) or descriptive (describing some occurrence or problem) in nature, descriptive statistical methods are often used in the analysis. However, in explanatory research (explaining an occurrence or problem), inferential statistical methods are used. This chapter presents statistical methods as well as a guide to selection of appropriate statistical methods in analyzing research data.

II. DATA ANALYSIS AND OVERVIEW OF STATISTICS

Research may not be considered complete without some sort of data analysis. Data analysis refers to applying statistical methods to the description of data and/or the inference of data in order to draw valid conclusions and make reasonable decisions on the basis of such analysis. Health science professionals must gain a conceptual understanding of statistics in order to perform data analysis as well as comprehend research literature. Some researchers mistakenly consider statistics a highly mathematical and difficult skill to master. Consequently, they are reluctant to learn this very useful tool. Be assured that you can have a good conceptual understanding of statistics without a good mathematical background. As a matter of fact, the only mathematical background needed for an understanding of this chapter is arithmetic.

It is true that if we do all the computations in any statistical analysis by hand or with calculators it can be tedious and often can result in errors. This is especially true with a large data set comprised of hundreds of observations. Fortunately, statistical packages (SPSSx, SAS, BMDP, Minitab, etc.) for both mainframe computers and microcomputers have become very popular in social science research. These packages enable researchers to perform many different types of data analysis in a simple and convenient manner. We strongly recommend that researchers take advantage of computers and statistical packages to perform their data analysis.* Researchers should focus on the selection of appropriate statistical procedures to analyze their data and/or interpret results.

A. DEFINITION OF STATISTICS

Perhaps the most complete definition of statistics is that given by Anderson and Bancroft[1] who describe it as "the science and art of the development and application of the most effective methods of collecting, tabulating, and interpreting quantitative data in such a manner that the fallibility of conclusions and estimates may be assessed by means of inductive reasoning based on the mathematics of probability".

The above definition may not be easily understood at first. Close examination, however, reveals that this definition describes two aspects of scientific problem solving: the description of observations, and inferences about scientific hypotheses from these observations.

These two aspects of scientific problem solving are in fact two major branches of statistics: descriptive and inferential. Descriptive statistics involves organizing, summarizing, and presenting data in a comprehensible format. Inferential statistics is often more important because it generalizes information from a sample to a larger populaiton.

It is important to understand that the application of any statistics will depend upon a number of factors, such as the level of measurement of variables (see Chapter 3), the number of variables, the design of the study, and the research hypothesis of the study. For example, if the measurement scales are at the nominal or ordinal level, nonparametric statistics (e.g., chi-square test), which does not require an underlying normal distribution for the data, may be used. When the measurement scales are at the interval or ratio level, parametric statistics (e.g., *t*-test, analysis of variance) may be used. As you are learning statistics, be sure to understand all the required antecedents that determine the chosen statistics.

* *Note:* Interested readers can obtain free statistical software by writing to M. Wang, Department of Health Education, Pennsylvania State University, University Park, PA 16802.

B. DESCRIPTIVE STATISTICS

Descriptive statistics is a method of obtaining indexes that summarize data to describe the major characteristics of a set of measures. In this section we will introduce commonly applied descriptive statistics: central tendency, variability, correlation, reliability, etc.

Central tendency and variability represent distributional characteristics of the variables in the study. Correlation identifies the relationship between two or more variables. Reliability reflects the consistency of the measures from a test instrument. All these descriptive measures help researchers select subsequent statistical analysis.

1. Measures of Central Tendency

A simple and convenient way of summarizing data is to find a single value that represents a set of measures. Since such values tend to lie centrally within a set of data, they are called measures of central tendency. The most commonly used central tendency measures are the mean, the median, and the mode.

a. Mean

The mean, also called the average or arithmetic average, is the most widely used measure of central tendency. The mean is computed by adding all individual scores for each case and dividing the sum by the number of cases. It is represented by the following formula:

$$\overline{X} = \frac{X_1 + X_2 + X_3 + \ldots + X_N}{N} = \frac{\sum X}{N}$$

where $\overline{X}$ = the mean, X = the score of each case in the data, and N = the total number of scores. For example, the mean of the numbers 5, 3, 8, 10, and 10 is

$$\frac{5 + 3 + 8 + 10 + 10}{5} = 7.2$$

The mean is appropriate *only* for interval or ratio data, not for nominal or ordinal data. For instance, if male and female categories are assigned values 1 and 2, respectively, the mean of male and female would have little relevance.

b. Median

The median is the middle point in a rank-ordered data set. When scores are arranged from highest to lowest, 50% of scores will be above this point and 50% will be below it.

Example 1: The set of numbers {3,3,5,7,8,9,10} has median = 7.

Example 2: The set of numbers {5,6,7,9,11,12,15,18} has median $\frac{1}{2}(9 + 11) = 10$.

The median is generally used for the interval level of measurement. It is also appropriate for the ordinal level when the number of categories is relatively large.

c. Mode

The mode is the value which occurs with the greatest frequency in a set of numbers. The mode may not exist in some data sets, while other data sets may have more than one mode and are called bimodal. Because of this instability the mode is not a recommended measure of central tendency.

Example 1: The data set {2,3,6,7,8,8,8,9,9,10,10} has mode 8.

Example 2: The data set {2,4,5,6,8,13} has no mode.

Example 3: The data set {1,2,3,3,3,4,5,6,6,6,8,8,9} has two modes, 3 and 6.

The mode can be used with variables at any level of measurement. However, the only appropriate measure of central tendency in a set of nominal data is the mode.

d. Choosing a Central Tendency Measure

The mean is an interval and ratio statistic; it is considered a more precise measure than the ordinal statistic, the median, and the nominal statistic, the mode. The mean takes into account the value of every score in a data set. Most parametric statistical tests (such as *t*-test, analysis of variance, etc.) may require the computation of the mean, but not the median or the mode. If the researcher is interested in the most frequently occurring score in a data set, the mode is clearly the best choice. However, the mode can be an extreme score in a data set and really not be the central measure. For example, through a weight control program ten adults lost weight as follows: 3, 3, 4, 5, 7, 9, 10, 11, 13, and 50 lbs. For these ten scores 3 is the mode, which represents only the two lowest values. The median is 8 because five scores fall above it and five scores fall below it; it is literally the middle score. The mean, which is 11.5, is higher than all but two of the scores in the data set because of the extreme score (50). Unlike the mean, the median will not be affected by the extreme scores in a data set.

2. Measures of Variability

Although a measure of central tendency provides us with useful information about the data distribution, it does not give us information about how the scores vary around the central tendency. Consider the following arbitrary scores for two groups of subjects:

Group A: 70 80 100 120 130 $\quad \overline{X} = \frac{500}{5} = 100$

Group B: 80 90 100 110 120 $\quad \overline{X} = \frac{500}{5} = 100$

By examining the mean one might conclude that Group A is no different than Group B. It is true that the two groups have the same average scores. However, careful examination of the scores reveals that the degree of dispersion or variability of these two groups is quite different. The scores in group A are obviously more widely spread around the mean than the scores in Group B. There is clearly a need for a measure that describes the data in terms of dispersion. The measure of the degree to which a set of data tends to spread about the mean value is called the variability or dispersion (these two terms are often used interchangeably). The most commonly used variation measures are range, variance, and standard deviation.

a. Range

The range of a set of data is the difference between the largest and smallest numbers in the data set. It is the simplest of all measures of variation. The range is suitable for variables at any level of measurement.

Example: The range of the set {2,3,5,5,8,12} is 12 − 2 = 10.

b. Variance

The variance, denoted as s^2 (sample) or σ^2 (population), is the appropriate measure of dispersion for interval or ratio level data. It computes how far each score is from the mean. Thus, the variance is one way of measuring how closely the individual scores in the data set cluster around the mean. This is done by

$$(X - \overline{X})$$

Each raw score will have a deviation score which is the distance between that raw score and the mean. Several deviation scores would not easily tell how far the raw scores are from the mean. The appropriate measure is to determine the average deviation from the mean. In this case, we would want to add all deviation scores together and divide by the total number of scores, just like we compute the mean. However, the sum of all deviation scores from the mean will always result in zero:

$$\sum (X - \overline{X}) = 0$$

Zero will obviously not reflect the degree of dispersion. To solve this problem we can square each deviation score and then add all squared scores together. Squaring eliminates the negative sign, so the sum of all deviation scores will always be positive. This is done by

$$\sum (X - \overline{X})^2$$

The result reflects the total amount of dispersion, which obviously depends on the number of scores. To adjust for the number of scores we should divide the result from the above formula by the total number of scores. This is done by the following formula:

$$\frac{\sum (X - \overline{X})^2}{N}$$

The resulting value is the variance. The variance is the average of the squared deviations from the mean of a set of scores. The variance of one data set can be compared with the variance of another data set even if two data sets may not have an equal number of scores. In reality, we will adjust the total number of scores by using N − 1 so that the formula for computing the variance is

$$\frac{\sum (X - \overline{X})^2}{(N - 1)}$$

The denominator (N − 1) provides an unbiased estimate of the population variance.[3] In practice, when the sample size is large this adjustment is trivial.

The above formula for computing variance is usually inconvenient because it involves obtaining the mean of a set of data. Another computational formula is

$$s^2 = \frac{\sum X^2}{N} - \left(\frac{\sum X}{N}\right)^2$$

The following computations illustrate how this formula is derived from the original formula. This illustration is only for the interested reader, and it is not necessary to know the computations.

$$s^2 = \frac{\sum (X - \bar{X})^2}{N} = \frac{\sum (X^2 - 2\bar{X}X + \bar{X}_2)}{N}$$

$$= \frac{\sum X_2 - 2\bar{X} \sum X + N\bar{X}^2}{N}$$

$$= \frac{\sum X^2}{N} - \frac{2\bar{X} \sum X}{N} + \bar{X}^2$$

$$= \frac{\sum X^2}{N} - 2\bar{X}^2 + \bar{X}^2$$

$$= \frac{\sum X^2}{N} - \bar{X}^2$$

$$= \frac{\sum X^2}{N} - \left(\frac{\sum X}{N}\right)^2$$

$$\text{or} \quad \sum X^2 - \frac{(\sum X)^2}{N}$$

Let us use a small data set to illustrate the computational processes with the above formula. For example, we wish to obtain the variance of a group of scores, 7,6,12,15,3,10,5,18.

$$\frac{\sum X^2}{N} = \frac{7^2 + 6^2 + 12^2 + 15^2 + 3^2 + 10^2 + 5^2 + 18^2}{8} = \frac{912}{8} = 114$$

$$\left(\frac{\sum X}{N}\right)^2 = \left(\frac{7 + 6 + 12 + 15 + 3 + 10 + 5 + 18}{8}\right)^2 = \left(\frac{76}{8}\right)^2 = 90.25$$

Then the variance is obtained as

$$s^2 = 114 - 90.25 = 23.75$$

The variance plays an important role in many statistical analyses that we will cover in later sections.

c. Standard Deviation

Because the variance involves the square of the deviation scores it does not reflect the original units of the measure. To use the same units to indicate the dispersion of a set of scores the standard deviation, denoted as s (sample) or σ (population), should be obtained. The standard deviation is simply the square root of the variance:

$$s = \sqrt{\frac{\sum (X - \bar{X})^2}{(N - 1)}}$$

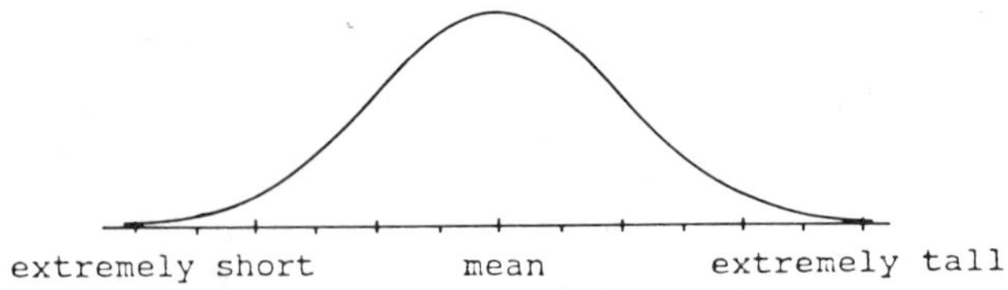

FIGURE 1. A hypothetical normal distribution of the heights of 10-year-old children.

With $s^2 = 23.75$ we obtain a standard deviation of

$$s = \sqrt{23.75} = 4.87$$

Now let us use the above-mentioned arbitrary scores from two groups of subjects.

Group A: 70 80 100 120 130
Group B: 80 90 100 110 120

Although the two groups have identical means (100), the standard deviation for Group A is larger than that of Group B (25.5 vs. 15.8).

The standard deviation (SD) is the most commonly used measure of dispersion. By knowing the standard deviation of a variable we can find out the expected frequency occurrence, assuming the data are distributed normally. This is accomplished by examining the table of the normal curve, which is included in Table 1 of the Appendix. For example, with a mean of 50 and a standard deviation of 10 a researcher desires to find the frequency occurrence of subjects scoring within ± 1.0 SD. The standard deviation value of 1.0 is found in the first column (entitled Z) of Table 1. To the right of this entry in the column 0.00 the number 0.3414, which is the frequency occurrence of 1 SD, is found. The number 0.3414 indicates that approximately 34% of the scores are associated with 1 SD. Thus, approximately 68% of the subjects scored between 40 and 60 (50 ± 10). The following section will introduce the basic concept of the normal curve.

C. THE NORMAL CURVE

The normal curve is a symmetric, bell-shaped curve. The distribution of many phenomena in nature tends to approximate this normal curve. For example, if we examine the height of 10-year-old children in the entire U.S., we will find that the majority of the children's heights cluster near the mean. As heights increase or decrease, the number of children decreases. Naturally, the further away from the mean, the fewer children we will find at each height. The extremely tall and the extremely short children would tend to be at the two ends. If we plot this curve it will resemble a normal curve (see Figure 1). The normal curve represents the theoretical frequency distribution, which can be expressed as follows:

$$y = \frac{1}{\sigma\sqrt{2\pi}} e^{-(X - \mu)^2/2\sigma^1}$$

where y = the frequency of any value of X, $\pi = 3.14159\ldots$, $e = 2.71828\ldots$, μ = the mean, and σ = the standard deviation.

This formula makes it possible to determine the ordinate of each value of X in the distribution. This operation, requiring considerable calculation, has been done by statisticians. Table 1 in the Appendix contains the areas of the normal curve that have been computed.

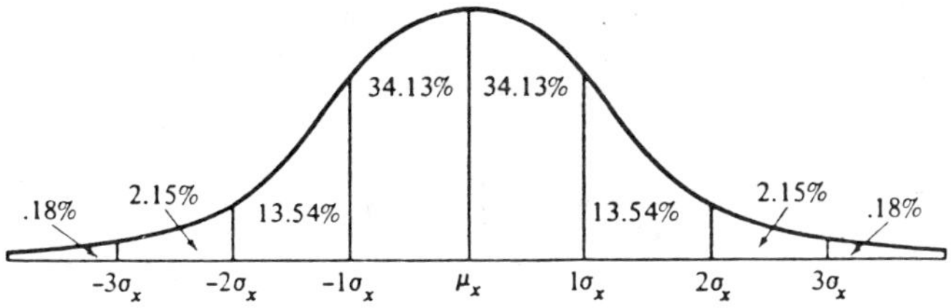

FIGURE 2. The normal curve.

The total area bounded by the curve and the X axis is 1. The mean of the normal curve is the point at which exactly 50% of the cases fall below and 50% fall above. Therefore, whenever data resemble the normal curve in distribution, we can obtain the frequency between the mean and any score expressed in terms of standard deviations from the mean. In other words, by knowing the standard deviation of a variable we can determine the percentage of the cases associated with the standard deviation. For example, approximately 34% of the cases in a normal distribution fall between the mean and 1 SD above or below the mean. If a boy has a height of 1 SD we can say this boy's height is greater than approximately 84% of all the boys measured. By examining the normal curve (see Figure 2) it can be seen that 1 SD covers the entire 50% of the curve on the left of the mean plus 34% on the right between the mean and the 1 SD. The approximate percentages of the areas associated with -3, -2, -1, 1, 2, and 3 SD above and below the mean have also been indicated in Figure 2. In theory approximately 68% of the scores in a normal distribution fall between ± 1 SD, 95% of the scores fall between ± 1.96 SD, and 99% of the scores fall between ± 2.58 SD. It is helpful to remember these percentages and the associated standard deviations since they are frequently referred to in statistical tests. The percentages above 3 and below -3 SD approach zero. These percentages can be interpreted as probabilities that are used for significance tests of statistics. Any score expressed in standard deviations from the mean is also a Z score, which will be discussed later.

Readers may wonder why it is that the normal curve is used rather than some other curve and also why researchers try to apply their data to a normal distribution. Actually, the normal curve was not invented by statisticians. It just happens that most biological traits, such as height and weight, follow the normal curve when measured on a large sample of human subjects and cast into a frequency distribution. Many other products of human effort and behavior, such as knowledge of personal hygiene or daily exercise time, are also distributed normally.

There are, however, some exceptions to this general rule. For example, when a measure of behavioral characteristics that is too easy or too difficult for the group tested is recorded, researchers do not expect a normal distribution. Suppose we present the knowledge statement "smoking is harmful to human lungs" to a group of adults, and the responses are "strongly disagree", "disagree", "undecided", "agree", and "strongly agree". We would expect that the majority of responses would be "strongly agree" or "agree". In this case, the distribution of responses would be skewed.

In some situations in which data have been obtained on subjects representing a mixture of two or more distinct groups with respect to the characteristics measured, a normal distribution is not expected. For example, if we were to measure the percentage of body fat of a group of college students composed of both males and females, the scores would likely have two peaks, which is a bimodal distribution (see Figure 3).

1. Skewness and Kurtosis

The normal curve is symmetric. Frequently, however, our data distributions, especially with small sample sizes, will show some degree of asymmetry, or departure from symmetry.

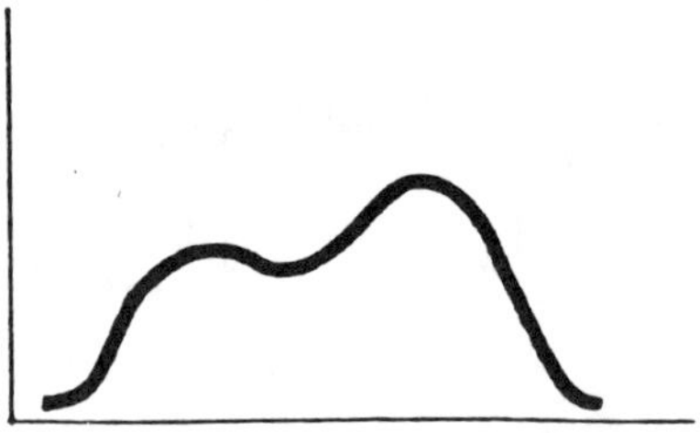

FIGURE 3. A hypothetical bimodal distribution of the percent of body fat of a group of college students.

A B

FIGURE 4. Skewed distributions. (A) Positively skewed; (B) negatively skewed.

Skewness is a statistic to measure the degree of asymmetry. If the distribution has a longer "tail" to the right of the peak than to the left, the distribution is skewed to the right or has positive skewness (see Figure 4A). If the reverse is true the distribution is said to be skewed to the left or to have a negative skewness (see Figure 4B). The value of skewness can be computed by

$$\text{Skewness} = \frac{\sum [(X - \overline{X}) / s]^3}{N}$$

where X is each individual score.

The value of skewness is zero when the distribution is a completely symmetric bell-shaped curve. A positive value indicates that the distribution is skewed to the right (i.e., positive skewness) and a negative value indicates that the distribution is skewed to the left (i.e., negative skewness).

While skewness describes the degree of symmetry of a distribution, kurtosis measures the height of a distribution curve. To compute kurtosis, use the formula

$$\text{Kurtosis} = \frac{\sum [(X - \overline{X}) / s]^4}{N} - 3$$

A positive kurtosis indicates that the distribution has a relatively high peak (Figure 5A); this is called leptokurtic. A negative kurtosis indicates that the distribution is relatively flat-topped (Figure 5B); this is called platykurtic. A normal distribution has a kurtosis of zero; this is called mesokurtic.

Skewness and kurtosis provide distributional information about the data. In statistical tests that assume a normal distribution of the data, skewness and kurtosis can be used to examine this assumption called normality.

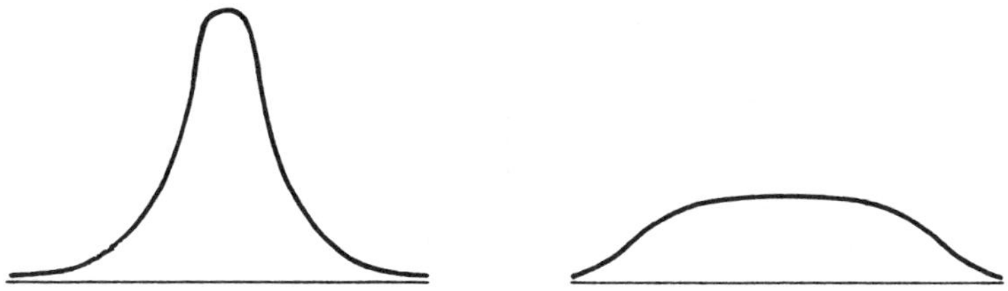

FIGURE 5. Non-normal kurtosis. (a) Leptokurtic; (b) platykurtic.

2. Standard Error

Assuming we could draw an infinite number of equally sized samples from a population and compute the mean of each sample, the means, which might not be identical, would form a normal distribution. The standard deviation of this distribution is called the standard error. The standard error provides us with the degree of discrepancy between the sample mean and the unknown population mean. The standard error, denoted as $\sigma_{\bar{x}}$, is not usuallly known. However, it can be estimated by dividing the standard deviation by the square root of the number of scores:

$$s_{\bar{X}} = \frac{s}{\sqrt{n}}$$

where $s_{\bar{x}}$, is the estimate of $\sigma_{\bar{x}}$. Using the above standard deviation as the example, the standard error is

$$s_{\bar{X}} = \frac{s}{\sqrt{N}} = \frac{4.87}{\sqrt{8}} = 1.72$$

The standard error is used in certain statistical tests and in computing confidence intervals, which will be discussed in Section II.C.4.

3. Z Scores

It is often difficult to compare the magnitude of variables measured in different units. For example, a person's income measured in the amount of dollars cannot be compared with this person's food intake measured in calories. However, we can transform scores into standard scores by standardizing the scale of measurement. The Z score transformation is a commonly used method to generate a new measure with a mean of 0 and a standard deviation of 1. This is done by

$$\text{Z score} = \frac{X - \bar{X}}{S}$$

where X is the original measure, $\bar{X}$ is the mean, and s is the standard deviation.

After the Z score transformation, each measure will be converted into a value equivalent to the number of standard deviations. In this sense, the standard deviation scores are standard scores or Z scores.

There are other types of standard scores (e.g., T-scores). Interested readers may want to read further.[18]

4. Confidence Interval

The topic of confidence intervals is classified as the theory of estimation of parameters. There are two commonly used estimations — the point estimation and the interval estimation.

When we consider a single value or number as the estimate of a parameter, it is called a point estimate. For example, a sample of 1000 households might be drawn randomly from the population of all households in a state for the purpose of estimating the average annual medical cost for the population. The mean, X, for the sample might equal \$235. This value, \$235, is a single point on the number line and, thus, provides a point estimate of the population mean, μ.

In contrast to a point estimate, an interval estimate consistes of an interval that we can say, with certain degrees of confidence, contains the parameter being estimated. This interval estimate can be obtained by the following formula:

$$\overline{X} \pm Z\sigma_{\overline{X}}$$

where Z is the Z score with a specified probability (e.g., Z of 1.96 is associated with a 95% confidence interval and Z of 2.58 is associated with a 99% confidence interval). $\sigma_{\bar{x}}$ is the standard error of the mean $\overline{X}$. For example, the average annual medical cost for 1000 randomly drawn households is \$235 with a standard deviation of \$50, assuming the annual medical costs are distributed normally and we want a 95% confidence interval for the annual mean medical cost of the population.

The point estimate of μ is $\overline{X}$ = \$235, the Z value corresponding to a confidence coefficient of 0.95 is 1.96, and the standard error of the estimate is

$$S_{\overline{X}} = \frac{S}{\sqrt{N}} = \frac{50}{\sqrt{1000}} = 1.58$$

Since the population from which the sample was drawn is distributed normally, we can obtain the confidence interval with the above formula:

$$235 \pm 1.96\,(1.58) = \{231.9 \sim 238.0\}$$

We are 95% confident that the mean of the population is contained in the interval 231.9 to 238.0. In other words, we know, if we were to repeat this sampling many times, that 95% of the intervals obtained in the above manner would contain the population mean.

The above formula in fact suggests that, all things being equal, the greater the standard error, the wider the confidence interval. The wider the confidence interval, the less accurate the estimate of the population parameter. Researchers should try to reduce the standard deviations of the measure.

D. CORRELATION

Both the measure of central tendency and that of variability are concerned with one single set of scores. Very often in our field a relationship exists between two (or more) variables. For example, blood pressure change may be related to age, heart rate may be associated with the aerobic exercise, weights of adults depend to some degree on their heights, and AIDS knowledge of the public may be related to the quality and the depth of educational programs. It is often desirable to express this relationship in a mathematical form that indicates the strength of this relationship. The statistical method for examining the relationship between two variables is called bivariate correlational analysis. The quantitative value of determining how strongly two variables are related is computed as a correlation coefficient, denoted as r. The relationship between two variables may be linear or

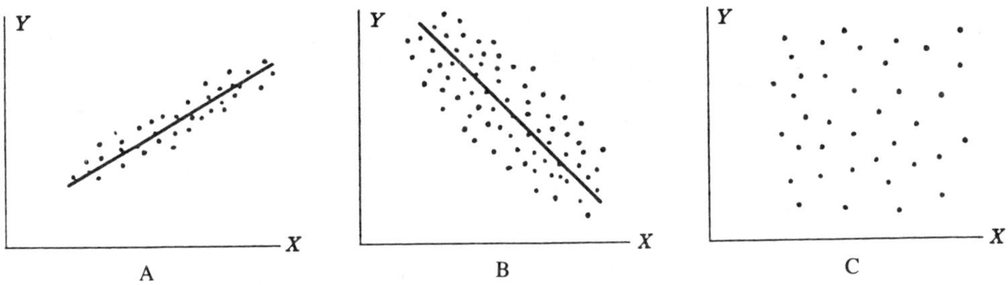

FIGURE 6. Scattergrams showing the relationships between variables X and Y. (A) Positive linear correlation; (B) negative linear correlation; (C) no correlation.

nonlinear. The nonlinear relationship is beyond the scope of this book. We will limit our discussion to only linear correlations.

In general, when two variables are said to be related or associated, either high scores on variable A are associated with high scores on variable B and low scores on variable A are related to low scores on variable B or high scores on variable A are associated with low scores variable B and low scores on variable A are related to high scores on variable B.

The correlation coefficient ranges from -1 to $+1$, representing both the direction and the strength of the relationship. Let X and Y denote two variables. A scatter diagram showing the location of two variables on a rectangular coordinate system will clearly illustrate both the degree and type of their relationship (see Figure 6).

In the scattergram, each pair of observations (usually, two measures have been made on a person or object or on the related persons or objects) is represented by a single point that is determined by the values of the observations. The distance on the X axis (horizontal line) represents the value of the observation of the X variable, and the distance on the Y axis (vertical line) represents the value of the observation of the Y variable. Figure 6 shows several scattergrams representing different types and degrees of association between two variables, X and Y. Generally speaking, a positive correlation is indicated when the trend of the dots is from lower left to upper right, while a negative correlation is present if the dots fall from upper left to lower right. No correlation is indicated when dots form a circular pattern. The more elliptical the pattern of dots becomes, the stronger the correlation between two variables.

The positive correlation, as shown in Figure 6A, suggests that a subject who gets a high score on variable A is likely to get a high score on variable B. Likewise, a subject who gets a low score on variable A is likely to get a low score on variable B. The negative correlation, as shown in Figure 6B, suggests that a subject who gets a high score on variable A is likely to get a low score on variable B, whereas a subject who gets a low score on variable A is likely to get a high score on variable B. In short, if Y tends to increase as X increases, as in Figure 6A, the relationship is positive. If Y tends to decrease as X increases, as in Figure 6B, the relationship is negative. When the change of Y does not depend on the change of X, as in Figure 6C, we say that there is no correlation between two variables.

Any correlation coefficient with a plus sign $(+)$ represents a positive relationship, any correlation coefficient with a minus sign $(-)$ represents a negative relationship, and 0 coefficient represents no relationship. The $+1$ represents a perfect positive correlation; the -1 represents a perfect negative correlation. In practice, it is extremely rare that we obtain correlation coefficients such as 0, $+1$, or -1.

When the coefficient approaches either $+1$ or -1, the relationship between two variables is considered strong. However, there is probably no absolute criterion about how great or how small a correlation coefficient should be to be considered strong or weak. The evaluation of the strength of a correlation coefficient should be based on both statistical value and practical significance.

TABLE 1
Correlation Coefficients and Their Significance

Absolute r	Relationship
1.0	Perfect
0.8	High
0.5	Moderate
0.3	Low
0.0	No

Statistical value refers to the size of the absolute value of r and its associated significant level (see Table 2 in the Appendix). Table 1 provides the rule of thumb for estimating the size of the coefficient. Readers may find different opinions in different statistics books.

Practical significance refers to the comparison of the obtained correlation coefficient with that of previous studies. For example, if the majority of previous studies found that the length of the smoking cessation training program was not strongly correlated with a reduced number of cigarettes (e.g., r was found to be below 0.4), but one of your studies found r to be 0.50, the finding could be considered practically significant and deserve further investigation. Even though $r = 0.50$ may not be considered very strong in terms of its absolute value, it is greater than the majority of previous findings.

Practical significance also refers to the comparison of the r value with what the theory suggests about two variables. This is especially true when there are few previous studies to be found. For example, theory suggests that there is a strong relationship between cognitive development and a child's age. Assume a researcher initiated a study and found $r = 0.50$ between the two variables. Based on the theory, the researcher would expect a greater correlation coefficient between the two variables. In this case, $r = 0.50$ may be considered low for the relationship between cognitive development and a child's age.

A word of caution is warranted here in interpreting the correlation coefficient. The correlation coefficient is only an index of the association between two variables. Such an association cannot be interpreted as a causal relationship. An association between two variables is a necessary, but not a sufficient, condition for causation.

In addition, readers should keep in mind that a correlation must not be interpreted as a percentage. The square of the correlation coefficient can indicate the percentage of one variable that is accounted for by the other variable. For example, the above correlation coefficient ($r = 0.50$) may suggest that 25% (0.50^2) of the variance along one variable can be accounted for by differences in another variable. The interpretation of r^2 will be discussed in Section II.D.3.

There are two common used bivariate correlation analyses: Pearson product-moment correlation and Spearman rank-order correlation.

1. Pearson Product-Moment Correlation

Pearson product-moment correlation is appropriate for two sets of interval or ratio level variables. The key assumption underlying the product-moment correlation is that the two variables are linear in form. To examine this assumption, plot a scattergram between two sets of scores (see Figure 6). This assumption is important in employing Pearson product-moment correlation analysis. The correlation coefficient may be misleading in nonlinear variables because two highly related variables in a nonlinear situation may result in a low coefficient.

TABLE 2
Sample Data of Variables X and Y

X	3	1	4	6	8	9	14	11
Y	2	1	4	4	5	7	9	8

TABLE 3
The Computational Demonstration of the Pearson Product-Moment Correlation

X	Y	$X - \overline{X}$	$Y - \overline{Y}$	$(X - \overline{X})(Y - \overline{Y})$	$(X - \overline{X})^2$	$(Y - \overline{Y})^2$
3	2	−4	−3	12	16	9
1	1	−6	−4	24	36	16
4	4	−3	−1	3	9	1
6	4	−1	−1	1	1	1
8	5	1	0	0	1	0
9	7	2	2	4	4	4
14	9	7	4	28	49	16
11	8	4	3	12	16	9
$\overline{X} = 7$	$\overline{Y} = 5$			84	132	56

The formula for computing Pearson product-moment correlation coefficient is

$$r = \frac{\sum (X - \overline{X})(Y - \overline{Y})}{\sqrt{\sum (X - \overline{X})^2 (Y - \overline{Y})^2}}$$

The above formula can also be written in the equivalent form for computational purposes:

$$r = \frac{N\sum XY - (\sum X)(\sum Y)}{\sqrt{[N \sum X^2 - (\sum X)^2][N \sum Y^2 - (\sum Y)^2]}}$$

Let us use a small data sample to illustrate the computational procedures. Suppose we have obtained two groups of scores, as in Table 2.

The computation work can be organized as in Table 3. The following formula is applied:

$$r = \frac{\sum (X - \overline{X})(Y - \overline{Y})}{\sqrt{\sum (X - \overline{X})^2 (Y - \overline{Y})^2}}$$

$$= \frac{84}{\sqrt{(132)(56)}} = 0.98$$

The correlation coefficient shows that there is a high correlation between the two variables. If we want to test the significance that the population coefficient of correlation, denoted by p, is estimated by the sample correlation coefficient r, we can apply the *t*-test (assuming this statistic involves a Student's distribution — see Section II.F.2.d):

$$t = \frac{r\sqrt{N - 2}}{\sqrt{1 - r^2}} = \frac{0.98\sqrt{8 - 2}}{\sqrt{1 - 0.98^2}} = 12.06$$

TABLE 4
Performance Ratings Assigned to Ten Work-Site Health Risk Reduction Programs by Two Experts

Subject	Rank 1	Rank 2	d	d^2
A	4	6	−2	4
B	3	3	0	0
C	6	8	−2	4
D	10	9	1	1
E	1	2	−1	1
F	2	1	1	1
G	8	10	−2	4
H	5	5	0	0
I	7	4	3	9
J	9	7	2	4
				$\Sigma d^2 = 28$

With N − 2 = 6 degrees of freedom (the concept of degrees of freedom will be discussed later), this *t* value is significant ($p < 0.001$; see Table 4 in the Appendix). More conveniently, the critical values of the Pearson correlation coefficient can be found directly in Table 2 in Appendix A.

2. Spearman Rank-Order Correlation

Spearman rank-order correlation is a nonparametric statistic for two sets of scores that are rank orders (i.e., ordinal data). The term "nonparametric" suggests that no assumptions are required about the distribution of scores on the variables. Rank-order correlation is designed to determine whether two rankings of the same cases are similar. For example, we might ask two health administration officials to rank order ten work sites according to the scope of their health promotion programs.

To compute the Spearman rank-order correlation the difference between each pair of ranks is computed and squared. The following formula is used for the Spearman rank-order correlation:

$$r = 1 - \frac{6 \sum d^2}{N(N^2 - 1)}$$

where d = the difference for each pair of ranks and N = number of paired observations. The test statistic is based on Student's distribution with degrees of freedom (df) = n − 2.

The following is an example. Two work site health experts were asked to rate the quality of implementation of health risk reduction programs of ten randomly selected work sites in Pennsylvania. The ratings are presented in Table 4. This table indicates the rank two experts assigned to each work site. One is the highest and ten the lowest value. The table also presents the values of difference between ranking, as well as the square of that difference. Based on the table results, the rank correlation coefficient is calculated as follows:

$$r = 1 - \frac{6(28)}{10(10^2 - 1)} = 0.83$$

The coefficient 0.83 indicates a positive relationship between the ranks of the ratings assigned to ten work site health risk reduction programs by two experts. The value of the coefficient is moderately high. We can test the significance of this correlation value using the t distribution as described above.

$$t = \frac{r\sqrt{N - 2)}}{\sqrt{1 - r^2}} = \frac{0.83\sqrt{10 - 2)}}{\sqrt{1 - 0.83^2}} = 4.20$$

Because the obtained t value (4.20) is greater than the critical value (2.31) we conclude that there is a significant positive relationship between the ratings of the two experts. Table 3 in the Appendix provides the critical values of the Spearman correlation coefficient.

3. r^2

It is easier to interpret the association between two variables by squaring the r (denoted by r^2, or R^2 in a multiple correlation situation) when we are more interested in the strength than the direction of an association. R^2 ranges from 0 to 1. It is the measure of the proportion of variance in one variable "explained" by other variable. Variance is the unknown nature of a variable represented by the variability of that variable. Variance is large when the observations are more spread out, and it is small when the observations are close to the mean. The objective of correlation analysis is to determine the extent to which variation in one variable is association with variation in another variable. Using the above Pearson correlation (0.98) as the example, the r^2 of 0.98 is 0.96. Thus, it can be said that 96% of the variation in variable X is reduced by introducing another variable, Y. Be cautioned again that r^2 or R^2 does not indicate a causal relationship.

4. Other Types of Correlations

a. Partial Correlation

Partial correlation provides the researcher with a single measure of the relationship between two variables while controlling for the effect of one or more additional variables. For example, the relationship between height and weight of a group of school children would be stronger when age varies, as opposed to the relationship of height and weight of a group of school children at a constant age. This is because as children grow older they become taller and heavier. Age is a factor that contributes to the strength of association between height and weight. Partial correlation allows a researcher to compute the correlation between height and weight while controlling for the variable age.

If we denote $r_{12.3}$ as the coefficient of partial correlation between X_1 and X_2 controlling for X_3, the formula to compute partial correlation is

$$r_{12.3} = \frac{r_{12} - r_{13}r_{23}}{\sqrt{1 - r_{13}^2)(1 - r_{23}^2)}}$$

When a correlation is computed without controlling for any other variable, we call it a zero-order correlation. When one variable is controlled in a partial correlation situation, we call it a first-order correlation. When two variables are controlled, then the correlation is called a second-order correlation.

b. Multiple Correlation

The multiple correlation, denoted as R, computes the correlation coefficient between one variable (usually a dependent variable) and two or more other variables (usually independent variables). The multiple correlation is useful in a multiple regression analysis. We will discuss multiple regression in Section II.G.

c. Point-Biserial Correlation

Point-biserial correlation is used when one variable is an interval or ratio score while another variable is dichotomous (e.g., gender, yes/no response). Interested readers can find computational procedures in Glass and Stanley's book.[6]

The following is the formula for computing the point-biserial correlation coefficient:

$$r_{pb} = \frac{\overline{X}_1 - \overline{X}_0}{s} \sqrt{\frac{n_1 n_0}{N\,(N-1)}}$$

where $\overline{X}_1$ is the mean on X of those who scored 1 on Y, $\overline{X}_0$ is the mean on X of those who scored 0 on Y, s is the standard deviation of all n scores on X, n_1 is the number of persons scoring 1 on Y, n_0 is the number of persons scoring 0 on Y, and $N = n_1 + n_0$.

d. Phi Coefficient

The phi coefficient is simply the Pearson product-moment correlationship for dichotomous data. The phi coefficient ranges from 0 to 1. Interested readers can find computational procedures and the interpretation of the phi coefficient in Glass and Stanley's book.[6]

$$\phi = \frac{P_{xy} - P_x P_y}{\sqrt{P_x Q_x P_y Q_y}}$$

where P_{xy} is the proportion of subjects scoring 1 on both X and Y, P_x is the proportion of subjects scoring 1 on X, P_y is the proportion of subjects scoring 1 on Y, Q_x is the proportion of subjects scoring 0 on X, and Q_y is the proportion of subjects scoring 0 on Y.

E. RELIABILITY IN DATA ANALYSIS

Results are considered useful only if they are repeatable. Reliability is the statistical technique that examines the extent to which results are repeatable. In practice there are many sources of measurement error that prevent the results from being repeatable — for example, inconsistent treatment conditions from subject to subject, changing environment during testing, subjects' fatigue, test items not understood, errors of scoring, data entry into the computer, etc. In this section we are particularly concerned with the reliability of survey instruments (e.g., questionnaires) that health science researchers frequently use.

The reliability of the survey instrument can be determined by two methods:

1. The responses of a group of randomly selected subjects are collected and then the consistency of responses to survey items of the same construct assessed. This is commonly called internal consistency reliability.
2. The responses of a group of randomly selected subjects are collected at one time; then the responses of the same group are collected at a later time, using the same survey instrument. The correlation between the two survey results is then assessed. This is called stability reliability, or test-retest reliability.

1. Internal Consistency Reliability

Internal consistency reliability is used to assess the survey items with a single construct such as knowledge, attitude, self-image, etc. The assumption underlying internal consistency is that items of the same construct (or trait) should produce similar responses. Internal consistency reliability is not applicable when the test instrument contains items that measure different traits. In this sense, we may interpret the reliability as an indicator of the test item homogeneity. A person who answers "yes" to the question "Do you agree that high schools should offer an AIDS education course?" would likely answer "yes" to the question "Do you agree that high schools should open an AIDS education clinic?" If the test instrument consists of items that measure different constructs, the reliability test of the whole instrument would likely show lower consistency than if individual constructs are examined. Very often a test instrument in the health science area consists of items that measure several dimensions. For example, a question asking "How do you rate your health condition?" may not have much bearing on the question "Do you think abortion should be banned in this country?" In this case, the reliability test should be carried out on each individual construct (in this example, health condition and opinion about abortion).

When items of a test instrument are measured at the interval level, the Cronbach alpha reliability coefficient[5] can be computed. When the test items consist of dichotomous responses (e.g., yes vs. no, agree vs. disagree, etc.), a reliability coefficient called KR-20[17] can be used. If responses are rank ordered (e.g., ordinal level), Kendall's coefficient of concordance[6] can be computed. Nominal data cannot be assessed for internal consistency reliability. All reliability coefficients range from zero to one. In practice, a negative reliability coefficient can be obtained when item responses do not show a consistent pattern. A coefficient of one indicates perfect reliability and zero indicates that measurements are all errors, although one and zero reliability coefficients rarely can be obtained.

2. Cronbach's Alpha Reliability Coefficient

Cronbach's alpha reliability applies to responses that have more than two options, with an interval measurement scale. For example, respondents may be asked to indicate how they agree to the statement "School medical facilities are out of date." The responses could be "strongly agree", "agree", "undecided", "disagree", and "strongly disagree". These five responses are assigned values 5, 4, 3, 2, and 1, respectively, in equal-appearing interval scales.

All measures are subject to errors. Each observed score has a true score component and an error score component. In other words, the variance of the observed scores is equal to the variance of true scores plus the variance of measurement errors. This relationship can be expressed in the following mathematical form:

$$\sigma_x^2 = \sigma_t^2 + \sigma_e^2$$

where σ_x^2 is the variance of the observed scores, σ_t^2 is the variance of the true scores, and σ_e^2 is the variance of the measurement errors.

The logic of Cronbach's alpha reliability is to find the proportion of the "true" variance measured to the observed variance in a survey instrument. The true variance is estimated by the observed variance minus the variance of measurement error. The formula for this definition is

$$\alpha = \frac{\sigma_x^2 - \sigma_e^2}{\sigma_x^2}$$

where α is the alpha reliability coefficient.

The computation for obtaining α is tedious. We will illustrate the computational procedures with a small number of items and a small number of subjects (Table 5). Researchers will probably have to employ computers and statistical packages to solve the problem. Readers are reminded that, in reality, the reliability test may not produce an accurate reliability coefficient with a small data set.

By using the analysis of variance it is possible to calculate the variance of true scores and the variance of measurement errors. Readers, however, should focus on the concept of reliability rather than the computations since they are subsidiary to the concept.

TABLE 5
Demonstration of Alpha Reliability Computations from a Hypothetical Example

Subject	Items 1	2	3	Row total
1	2	2	1	5
2	3	2	3	8
3	2	2	2	6
4	2	2	1	5
5	1	1	1	3
Column total	10	9	8	$\Sigma X = 27$
				$(\Sigma X)^2 = 729$
				$\Sigma X^2 = 55$

1. Obtaining sum of squares between items (SSb):

$$SSb = \frac{10^2 + 9^2 + 8^2}{5} - \frac{729}{15} = \frac{245}{5} - \frac{729}{15} = 0.4$$

2. Obtaining sum of squares within individuals (SSw):

$$SSw = \frac{5^2 + 8^2 + 6^2 + 5^2 + 3^2}{3} - \frac{729}{15} = \frac{159}{3} - \frac{729}{15} = 4.4$$

3. Obtaining sum of squares error (SSe)

$$SSe = 55 - \frac{245}{5} - \frac{159}{3} + \frac{729}{15} = 1.6$$

Mean squares (MS) are obtained by dividing the sum of squares by the associated degrees of freedom.

Source	SS	df	MS
Items	0.4	2	0.2
Individual	4.4	4	<u>1.1</u>
Error	1.6	8	<u>0.2</u>

Our interest lies in the variance among individuals and the variance of measurement errors, which are underlined. The mean square among individuals is the estimated observed variance, while the mean square error is the estimated variance of measurement error. Applying the alpha reliability formula we obtain the reliability coefficient:

$$\alpha = (1.1 - 0.2) / 1.1 = 0.82$$

Through these computation procedures we attain some idea of what would contribute to a high reliability and what would contribute to a low reliability. The smaller the difference among items totals (i.e., the sum of column totals), the lower the reliability coefficient. The greater the difference among individuals totals (i.e., the sum of row totals), the higher the reliability coefficient. Let us use the above example to illustrate this.

If we change the fourth subject's score from 2 2 1 to 1 2 2, the row totals do not change, but the column totals become 9 9 9.

Subject	Items 1	2	3	Row total
1	2	2	1	5
2	3	2	3	8
3	2	2	2	6
4	1	2	2	5
5	1	1	1	3
Column total	9	9	9	$\Sigma X = 27$ $(\Sigma X)^2 = 729$ $\Sigma X^2 = 55$

$$SSb = \frac{9^2 + 9^2 + 9^2}{5} - \frac{729}{15} = \frac{243}{5} - \frac{729}{15} = 0$$

$$SSw = 4.4 \text{ (no change)}$$

$$SSe = 55 - \frac{243}{5} - \frac{159}{3} + \frac{729}{15} = 2.0$$

Source	SS	df	MS
Items	0	2	0
Individual	4.4	4	1.1
Error	2.0	8	0.25

$$\alpha = (1.1 - 0.25) / 1.1 = 0.77$$

It is seen that the alpha reliability coefficient dropped from 0.82 to 0.77 due to the column totals change.

We can also change the row values without changing the column totals. If we switch the values of the third item of subjects 1 and 2 to change the rows 1 and 2 totals from 5 and 8, respectively, to 7 and 6 and then compute the reliability coefficient, the alpha coefficient decreased from 0.82 to 0.54.

Subject	Items 1	2	3	Row total
1	2	2	3	7
2	3	2	1	6
3	2	2	2	6
4	2	2	1	5
5	1	1	1	3
Column Total	10	9	8	$\Sigma X = 27$ $(\Sigma X)^2 = 729$ $\Sigma X^2 = 55$

$$SSb = 0.4 \text{ (no change)}$$

$$SSw = \frac{7^2 + 6^2 + 6^2 + 5^2 + 3^2}{3} - \frac{927}{15} = \frac{155}{3} - \frac{927}{15} = 3.1$$

$$SSe = 55 - \frac{245}{5} - \frac{155}{3} + \frac{729}{15} = 2.9$$

Source	SS	df	MS
Items	0.4	2	0.2
Individual	3.1	4	0.78
Error	2.9	8	0.36

$$\alpha = (0.78 - 0.36) / 0.78 = 0.54$$

3. KR-20 Reliability Coefficient

When the responses of test items are dichotomous, a formula developed by Kuder and Richardson[17] to measure reliability, known as KR-20, is used.

$$\text{KR-20} = \left(\frac{n}{n - 1}\right)\left(\frac{s^2 - pq}{s^2}\right)$$

where n = the number of items in a test instrument, s^2 = the variance of the test scores, and p = the proportion of the correct answers to a given question $q = 1 - p$.

An even simpler formula, also developed by Kuder and Richardson, is known as KR-21.

$$\text{KR-21} = \frac{ns^2 - M(n - M)}{(n - 1)s^2}$$

where M is the mean of the test scores.

Let us use the following example. Suppose we obtained $s^2 = 8.05$ and mean $= 7.05$ from a ten-item instrument. Applying the KR-21 formula,

$$\text{KR-21} = \frac{(10)\,(8.05) - 7.05(10 - 7.05)}{(10 - 1)(8.05)} = 0.82$$

so the obtained KR-21 is 0.82.

4. Split-Half Method

In the split-half method a single survey instrument consists of two parts in parallel forms, each of which measures the same construct. Thus, we can correlate the scores of one half of the items with scores of the other half of the items of the instrument. The formula for computing the split-half reliability is

$$\text{split-half reliability} = \frac{2r_{12}}{1 + r_{12}}$$

where r_{12} is the correlation between the two split samples.

TABLE 6
An Example of the Relationship between the Number of Items and the Reliability Coefficient

Number of items	α-Coefficient
15	0.82
12	0.78
10	0.72
8	0.65
6	0.57
4	0.44

Suppose the correlation coefficient between two split items is 0.87. Applying the preceding formula,

$$\text{Split-half reliability} = \frac{2 \times 0.87}{1 + 0.87} = 0.93$$

There are a number of ways to split half of the test items. For example, we can split the instrument by dividing the items into odd and even items or by randomly assigning the total number of items to one or the other of two half-tests. Any division may be legitimate as long as the assigned numbers of the original items do not fall into any logical order.

5. Interpretation of Reliability Coefficient

Students who develop a survey instrument or apply an instrument developed by other researchers often ask the question ''How big must the reliability coefficient be to be considered reliable?'' It is a good question, but there is no easy answer. As previously discussed relative to correlation coefficients, reliability depends on statistical as well as practical importance. The following points may help readers understand reliability coefficients.

1. As a general rule, an alpha coefficient that exceeds 0.9 may be considered high; 0.8, moderately high; and 0.7, acceptable. Below 0.6 may be considered low. These criteria will depend on the nature of the test instrument, however, as well as the number of items in the test instrument, the type of measurement scale, etc.
2. A test instrument with abstract questions or hard-to-quantify measures tends to produce a lower reliability coefficient. Instruments examining knowledge and attitude scales are likely to result in reliability coefficients above 0.8.
3. A longer test instrument tends to have a higher reliability coefficient than a shorter instrument. In a 15-item questionnaire examining AIDS knowledge a reliability coefficient of 0.82 was found. From these 15 items we randomly selected 12, 10, 8, 6, and 4 items to examine how the number of items would affect the reliability coefficient. Results are presented Table 6. We can see from this table that as the number of items decreased there was a corresponding decrease in alpha reliability.
4. The average correlation among items determines the reliability coefficient. If the average correlation is zero or near zero we would consider the items as a group having no common core, and it is not sensible to consider them as measuring a single attribute. Item-total analysis may be performed to examine individual items for their contribution to the reliability coefficient.
5. A test instrument with multiple option responses (e.g., strongly agree, agree, undecided, disagree, and strongly disagree) tends to produce higher reliability coefficients

TABLE 7
An Example of Item-Total Analysis

Items	Item total r*	α if deleted
1	0.53	0.85
2	0.27	0.87
3	0.65	0.84
4	0.26	0.86
5	0.73	0.83
6	0.52	0.85
7	0.81	0.82
8	0.65	0.84
9	0.55	0.85
10	0.60	0.84

Note: When item-total correlation is computed, the total score is the total score from all items except the item that is being correlated. Some statistical packages call this item-total correlation the "corrected item-total correlation".[19]

than dichotomous responses (e.g., yes/no, agree/disagree). Nunnally[15] suggests that 30 dichotomous items have to be used to achieve the reliability of 20 multiple-choice items in a similar content situation.

6. As a good rule of thumb, the number of subjects used for reliability should be at least ten times the number of items. This rule of thumb may not be practical if the items number is too high (i.e., 70 and more). In any case, five subjects per item should be considered a bare minimum.[15]

When the coefficient α proves to be low, it may suggest that either the test is too short or some items may not have a common construct with most of the items. The solution may be to reconstruct some of the items and/or add more items to the test, then perform an item analysis.

6. Item Analysis

There are two major ways to perform an item analysis: percentage response and item-total correlation.

1. Percentage response. Suppose we have a knowledge item with four response choices on a difficult continuum. We would expect that the percentage of a response to each choice would be 25% or 0.25 probability with a large, randomly selected group of subjects. If we should find an extreme percentage response on either end of the choices (e.g., almost every subject selected the correct answer or almost every subject selected the wrong response), it would indicate that this item does not discriminate among subjects and, thus, does not contribute to the reliability of the test.
2. Item-total correlation. Item-total correlations and corrected α may be computed for the purpose of evaluating items when computer and statistical packages are available. If an item measures a common construct with a number of items, then this item should be related to the other items as a whole. Item-total analysis computes the correlation between each individual test item and all the remaining items as a group.

Table 7 presents an example of results of item-total analysis. The items that correlate highly with total scores are better items than those items having relatively low correlations with total scores. The items with low item-total correlations will lower the alpha reliability. This fact can be seen in the last column, ''alpha if deleted'', which is the alpha value of the test instrument when this item is deleted. Again, there is no absolute criterion to determine what item-total correlation coefficient will be discarded or retained. As a rule of thumb, items with an item-total correlation lower than 0.3 may not be good and should be discarded.

F. INFERENTIAL STATISTICS

Inferential statistics is the process of making decisions about populations on the basis of sample information. Such decisions are called statistical decisions. For example, we may wish to decide, based on the sample data, where one health promotion program is better than another in reducing health risk scores or whether a new drug is really effective in curing a disease, etc. Statistical inference is the inductive reasoning that is used in the statistical testing of a hypothesis. The theoretical basis for such statistical testing is the sampling distribution.

1. Testing a Hypothesis Using Sampling Distribution

When we make inferences about a population based on the sample information, we want to know with what accuracy a sample value, called statistics, estimates the population characteristics, called parameters. Statistics are denoted by English letters, such as $\overline{X}$ (mean) and s (standard deviation), while parameters are denoted by Greek letters, such as μ (mean) and σ (standard deviation). The value of a statistic generally varies from one sample to another, even though the samples are all drawn randomly from the same population. For example, we draw a random sample of 100 from a large population and obtain the statistic (i.e., mean). Suppose we repeat this process hundreds of times and compute the mean from each sample. Since these means vary, a new frequency distribution can be established for them. This distribution is called a sample distribution. It is the frequency distribution of a hypothetically infinite number of statistics computed from separate random samples of the same size from the same population. The expected value of any statistic is the mean of the sampling distribution. The standard error (which we discussed previously) of any statistic is the standard deviation of this sampling distribution. Thus, by obtaining a value of the statistic of a random sample, this sample statistic can be compared to the sampling distribution and the chances (or probability) of the sample being drawn from the population can be determined. By the same token, we can draw more than one random sample and determine the chances of the sample being drawn from the same population. In addition, a sampling distribution can be used to estimate unknown population parameters from a known sample value and state the degree of confidence of the estimation. Sampling distribution is also useful in determining whether observed differences between two samples are really significant or are due to chance variation only. The reader who is interested in an advanced discussion in this topic should read the book by Glass and Stanley.[6]

2. Statistical Hypothesis Testing

In making statistical decisions it is useful to speculate about the populations involved. This is called a statistical hypothesis, a statement about an unknown parameter to be tested in the population. A common research approach is to state the possible outcomes as no difference between specified population parameters. For example, if we wish to decide if the experimental group scores higher than the control group, we formulate the hypothesis that there is no difference between the experimental and control groups. This is called the null hypothesis and is denoted by H_0. Of course, this is not usually the anticipated outcome of the study. Hence, it is also common to formulate an alternative hypothesis for each null hypothesis, denoted by H_1, stating that one group differs from another. For example, with the above null hypothesis we may state that the experimental group scores higher than the

		The researcher's decision about the null hypothesis	
		true	false
Null hypothesis in fact is	true	right	Type I
	false	Type II	right

FIGURE 7. Schematic representation of Type I and Type II errors.

control group as the alternative hypothesis, or we may simply state that the two groups differ. In this case, the null hypothesis will be

$$H_0\text{: Group 1} = \text{Group 2}$$

$$H_1\text{: Group 1} > \text{Group 2 or Group 1} \neq \text{Group 2}$$

It is important to note that the null hypothesis is a statement about population parameters. Therefore, the hypothesis should be stated in population parameters:

$$H_0\text{: } \mu_1 = \mu_2$$

$$H_1\text{: } \mu_1 > \mu_2 \text{ or } \mu_1 \neq \mu_2$$

When the researcher conducts the experiment, he or she has reason to believe that certain treatments (usually independent variables) may influence a particular outcome (usually dependent variables). However, the decision is based on the results of hypothesis testing.

Once hypotheses are formulated we may obtain data and perform some type of statistics. If statistical results show that the experimental group scores substantially higher than the value expected for the control group from the probability of the sampling distribution, we would reject the null hypothesis and conclude that there is a significant difference between the two groups or that the experimental group scores significantly higher than the control group. Our conclusions will always be subject to some error because each sample drawn from a given population is expected to differ to some degree from the population. This is called sampling error, which is simply the difference between the value of a sample statistic and the corresponding population parameter. While a variety of procedures can be used to minimize sampling error (see sampling discussion in Chapter 3 of this book), no procedure can eliminate sampling error completely. That is to say, it is not possible to reject a null hypothesis with complete certainty, and error is always a possibility.

a. *Type I and Type II Errors*

When the decision is to reject a null hypothesis, but in fact the null hypothesis should be accepted, a Type I error is made. On the other hand, if the decision is to accept a null hypothesis when it should be rejected, a Type II error is made (see Figure 7). In either case a wrong decision has been made. It is obvious that in testing a hypothesis it is important to minimize chances of making wrong decisions. This is not a simple matter because an attempt to reduce one type of error is accompanied by an increase in the other type of error. In practice, the researcher should weigh the relative consequences of both errors so that a rule of decision will be made to reduce the type of error with the more serious consequence.

In general, most researchers consider the consequences of a Type I error more serious than the consequences of a Type II error, since a significant treatment may lead to a specific action. For example, if a new drug is found to be more effective than an existing drug and the premise is not true, a Type I error is made. The drug company may allocate large sums of money to produce the new drug that is no more effective than the original drug. However, if the researcher makes the decision that the new drug does not have any better effect than the existing one, but in reality there is an improvement, a Type II error is made. No action occurs as a result of the study and the status quo is maintained as before the experiment started. It is likely that this study may be carried out again some time in the future by the same researcher or a different one. The significant effect may be found and the action may take place at this time. The consequence of a Type II error due to the first experiment only results in delaying the action rather than taking an unnecessary action, as is the consequence of a Type I error.

b. Level of Significance

When we test a statistical hypothesis we draw a conclusion about the probability that it is correct, rather than statements of absolute truth. Investigators have to decide beforehand what probability of error they want to risk. This predetermined maximum probability the researcher is willing to risk of a Type I error is called the level of significance. The level of significance, denoted by α, should be determined before conducting the study so that the results obtained will not influence the researcher's choice.

The level of significance is an arbitrary criterion. In practice, a level of significance of 0.05 or 0.01 is customary for social science research. Assuming a researcher selects α = 0.05 level of significance in a test of a null hypothesis, there are about 5 chances in 100 that the researcher would reject the null hypothesis when it should be accepted; that is to say, the researcher is 95% confident that the right decision will be made. In such cases we say the hypothesis is rejected at a 0.05 level of significance.

The statistical value (i.e., from statistical tables) associated with the level of significance is called the critical value. Only when the obtained statistical value is greater than the critical value can we reject the null hypothesis. We indicate that the probability of the null hypothesis being true is less than the predetermined acceptable probability ($p < 0.05$ or $p < 0.01$), and the results are significant. Otherwise, we declare that the results are statistically nonsignificant, and the null hypothesis is accepted. Readers are cautioned that the statistical term *significant* only refers to the probability of chance being less than the predetermined level of significance; it does not suggest how important the observed difference is. Results of studies can be statistically significant without practical importance. For example, a pharmacy plans to produce a new drug to help people lose weight. The cost for this product is estimated at $10 million. The company first produced 1000 samples to test the effect of this drug and to decide if large-scale production was necessary. The test showed that the improvement of the new drug was only trivial compared to one existing drug. This miniscule improvement, however, may be statistically significant due to the fact that a large sample (1000) was tested. In this case it may be legitimate for the researcher to conclude that, in consideration of the substantial estimated cost for this project, the marginal improvement may be considered nonsignificant.

c. Steps of Testing Statistical Hypotheses

Before we discuss different inferential statistical tests we will briefly review the steps involved.

Step 1 — State the null and alternative hypotheses which will provide direction to sample selection and to the statistical analysis.

Step 2 — Identify the population and determine the sampling procedures and sample size. Sample size should provide enough statistical power, which is discussed in Chapter 3.

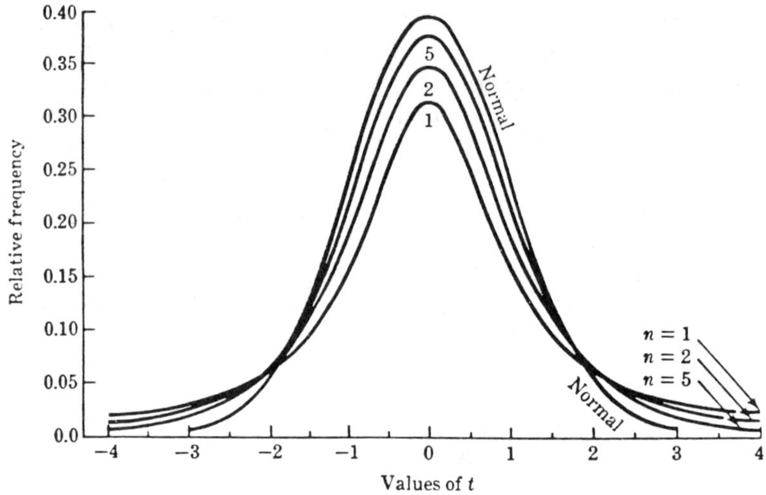

FIGURE 8. The t-distribution for various sample sizes.

Step 3 — Determine the level of significance and the appropriate statistical procedures. They will depend on the measurement level of variables, the number of independent and dependent variables, the design of the study, etc.

Step 4 — Compare the statistics value with the critical value from statistic tables. Decide if the null hypothesis will be rejected or accepted.

There are numerous methods of testing null hypotheses. We will introduce the most widely used in the following section.

d. t-test

The t-test is used to test the difference between two means when the measurement is at the interval or ratio level. The statistical inference, using the normal curve as a model for the sampling distribution, is appropriate for large samples. When the sample size is small, especially smaller than 30, using the normal curve as a probability model is inadequate. A British statistician, William S. Gossett, developed a series of different sampling distributions called t-distributions (also known as the Student's t-distribution because of Gossett's pen name) for testing hypotheses using sample samples.[18] When the sample size is infinite the t-distribution is identical to normal curve. When the sample size is greater than 30 the t-distribution is close to normal. When the sample size becomes smaller the t-distribution becomes greater in the tail area than the normal distribution (see Figure 8). The t-distribution indicates that with smaller samples it is more difficult to reject a null hypothesis. We do not have to know how to calculate t-distributions with different sample sizes. There is a statistical table of t values for different sample sizes, indicated by the degrees of freedom (see Table 4 in the Appendix).

By definition, the number of degrees of freedom (df) of a statistic is the number of observations free to vary around a constant parameter. This is obtained by subtracting one from the number of independent observations in the sample.

There are three applications to t-test: (1) to test a hypothesis about the mean, (2) to compare the means of two independent groups, and (3) to compare the means of a paired group.

i. The t-Test of a Hypothesis About the Mean

The t-test can be used to test a hypothesis about the mean of a normally distributed population. A hypothesized mean will be compared with the mean of the population under study. The t-test is employed to see if a sample mean drawn from the population differs significantly from the norm group. For example, a sample of ten children's IQ scores was obtained as 101, 99, 120, 79, 111, 98, 106, 112, 87, and 97. The population mean is 100 and the standard deviation is 10. The t-test will determine whether these ten children are from a different population than the norm which was used to standardize the measuring instrument. The t value is calculated by:

$$t = \frac{\overline{X} - \mu_{\overline{J}}}{S_{\overline{X}}} = \frac{\overline{X} - \mu_{\overline{J}}}{S/\sqrt{N}}$$

where $\overline{X}$ is the mean of a sample and μ is the mean of the population. The s is the standard deviation of the mean.

If the calculated t value is greater than the critical value, we reject the null hypothesis and state that the difference between two means is significant at the specified level (e.g., 0.05). If the calculated t value is smaller than the critical value, we retain the null hypothesis and state that the difference between two means is not significant. With the above scores we obtain $t = 0.3$. Consulting with Table 4 in the Appendix, we locate df $= (10 - 1) = 9$ in the first column and find the entry value of 2.26 under the column labeled 0.05. The value, 2.26, is the critical value. Our obtained t value (0.3) is smaller than the critical value ($p > 0.05$). We retain the null hypothesis and conclude that there is no significant difference between the mean IQ scores of ten students and the population mean.

This t-test is analogous to using the Z score,

$$Z = \frac{\overline{X} - \mu}{s/\sqrt{N - 1}}$$

The difference is that Z is distributed normally while t follows the Student's t-distribution. When the sample size is large, these two statistics tend to agree.

ii. The t-Test for Two Independent Samples

Experimental designs that select two random samples from a population and assign one group to an experimental treatment and another group to a control condition are popular. Generally, difference can be observed between the two groups. The researcher wants to know whether this observed difference is statistically significant or if the difference may be attributable to chance only. The t-test can be used to examine whether the treatment may cause significant differences between the two groups. This t-test is called a t-test for independent samples. The term *independent* refers to the fact that the groups are neither matched nor paired, nor does any other relationship exist between them. The subjects in the two groups for the independent t-test must be randomly selected and assigned to treatment or control groups. In this case, the significant difference observed between two groups may be attributable to the treatment effect.

Another common situation is when two random groups are selected from two different populations. The t-test can be used to examine the existing differences between the two groups. In this case the two groups are often selected for certain characteristics, such as sex, age, or other background information (e.g., education, occupation, income, experience, etc.). Independent variables are sometimes referred to as differential variables. Experimental treatments will not be given to the groups. The question asked is whether these two groups, selected by different characteristics, are from the same population or different populations.

Let us illustrate the independent t-test with an example. Suppose 14 girls and 16 boys were selected randomly from one school to test their systolic blood pressure. The 14 girls showed a mean of 112 with a standard deviation of 8, while the 16 boys showed a mean of 107 with a standard deviation of 10. Is there a significant difference at the 0.05 level (assuming the homogeneity requirement is met) between the systolic blood pressure of boys and girls?

Using μ_1 and μ_2 to denote the population mean systolic blood pressure of girls and boys, respectively, we state two hypotheses:

- H_0: $\mu_1 = \mu_2$, and there is no real difference between two groups.
- H_1: $\mu_1 \neq \mu_2$, and there is a significant difference between two groups.

There are two assumptions associated with the independent *t*-test:

1. Normality — the measures in both samples are distributed normally. This assumption can be examined using either the histogram of data distribution or skewness and kurtosis, as discussed previously.
2. Homogeneity — the variances of the two populations are equal. This assumption can be examined by the Hartley's F_{max} test,[9] which is used widely. The F_{max} test computes the ratio

$$F_{max} = \frac{\text{larger } s^2}{\text{smaller } s^2}$$

If the F_{max} value is greater than the critical value provided by Table 5 in the Appendix, the assumption ($\sigma_1^2 = \sigma_2^2$) is violated; otherwise, the assumption is met. Strictly speaking, the F_{max} test requires that two samples have equal n values. If they do not, the average number of degrees of freedom should be used. The number of degrees of freedom is denoted as v in Table 5 in the Appendix. For example, if one group has eight subjects and another group has six subjects, df = (8 − 1 + 6 − 1)/2 = 6.

When the homogeneity assumption is met the *t* is computed:

$$t = \frac{\overline{X}_1 - \overline{X}_2}{s_{\overline{d}}}$$

However, if the homogeneity assumption is violated, the *t* statistic is not distributed as a Student's *t*. The probability for *t* can be approximated by treating it as *t*, with the degrees of freedom computed as follows:

$$df = \frac{[(s_1^2/n_1) + (s_2^2/n_2)]^2}{[(s_1^2/n_1)^2/(n_1 - 1)] + [(s_2^2/n_2)^2/(n_2 - 1)]}$$

Experienced researchers know that tests of normality and homogeneity assumptions are very effective with small samples, and violations of these assumptions with small samples are likely to be detected. However, strong mathematical and empirical evidence has shown that a certain degree of violation of assumptions will not seriously affect the statistical results. This effect is called the "robustness" of statistical procedures. The reader needs to bear in mind that when both assumptions are violated there is a higher chance of making a Type I error — finding a significant difference when in fact there is no significant difference. In our sample data, F_{max} is computed as follows:

$$F_{max} = \frac{10^2}{8^2} = \frac{100}{64} = 1.5$$

The degrees of freedom are $(16 - 1 + 14 - 1)/2 = 15$. Examining Table 5 in the Appendix we find the value of 15 in the first column labeled v, and the critical value for two groups is 2.86 in the second column labeled 2. (Note: the value 2 indicates the number of groups, and F_{max} can examine more than two groups.) Our obtained F_{max} is smaller than 2.86; hence, the assumption is met.

Applying the *t*-test formula,

$$t = \frac{\overline{X}_1 - \overline{X}_2}{s_{\bar{d}}}$$

where

$$S_{\bar{d}} = \sqrt{s_1^2/n_1 + s_2^2/n_2}$$

We compute

$$s_{\bar{d}} = \sqrt{(8^2 / 14) + (10^2 / 16)} = 3.289$$

We then compute

$$t = \frac{112 - 107}{3.289} = 1.52$$

With the degrees of freedom, $N_1 + N_2 - 2 = 16 + 14 - 2 = 28$, the critical value is 2.048 (see Table 4 in the Appendix) for the 0.05 level of significance. Our computed *t* value, 1.52, is smaller than this value; thus, we cannot reject H_0 at a 0.05 level of significance. We conclude that there is no significant difference between the systolic blood pressure of boys and girls ($p > 0.05$). Note: If the variances of the groups differ, we should compute the $s_{\bar{d}}$ by the formula

$$\sqrt{\frac{(n_1 - 1)s_1^2 + (n_2 - 1)s_2^2}{n_1 + n_2 - 2}\left(\frac{1}{n_1} + \frac{1}{n_2}\right)}$$

iii. The t-Test for Two Paired Samples

In the independent samples each subject is randomly selected from the population, and formation of one group has no bearing on the formation of another group. Randomization in this design is the only provision for controlling the irrelevant variables. There is no good way, however, to estimate the effect of the subjects' variation from group to group that may affect the dependent measures. There is always a question regarding the precision of an experiment.

If the irrelevant variables due to subjects' variation can be controlled, the precision of the experiment will be increased. This can be achieved by using two related samples. For example, we may wish to compare scores of a group of the same subjects measured before and after a diet program. Also, in certain situations we wish to match the subjects of two groups based on their similar qualities. In this case, the independent *t*-test would not be appropriate because the dependent variable would likely be correlated. Therefore, the paired *t*-test (sometimes called the correlated *t*-test) must be used.

Let us look at one example. An investigator is interested in finding out how jogging might reduce a normal, healthy person's diastolic blood pressure. To investigate this question,

TABLE 8
The Diastolic Blood Pressure Scores of Eight Subjects Before and After 3 Months of Jogging

Subject	Before (X_1)	After (X_2)	d ($X_1 - X_2$)	d^2
1	66	61	5	25
2	80	76	4	16
3	77	72	5	25
4	74	68	6	36
5	82	76	6	36
6	73	68	5	25
7	87	84	3	9
8	69	68	1	1
			35	173
			mean 4.375	

eight healthy subjects from a community went through a 3-month jogging program. Their diastolic blood pressure measurements were obtained immediately before and after the 3 months of jogging. Their scores are presented in Tablc 8.

The assumption of the paired *t*-test is that the two measures (before and after 3 months of jogging) must be positively related; if the relationship is negative, the initial assumption of using the paired *t*-test should be reexamined. In this case, the correlation between the two measures is positive (r = 0.97); the assumption is met.

The paired *t*-test formula is as follows:

$$t = \frac{\bar{d}}{s_{\bar{d}}}$$

where t = the t value for the paired *t*-test, d = the difference between the paired scores, $\bar{d}$ = the mean of the difference, N = the number of pairs, and $s_{\bar{d}}$ = the sum of squared difference scores. $s_{\bar{d}}$ is computed as

$$s_{\bar{d}} = \sum \frac{d^2 - (\sum d)^2 / N}{N(N - 1)}$$

$$s_{\bar{d}} = \frac{173 - 35 / 8}{8(8 - 1)} = 0.357$$

Our computed paired t is

$$t = \frac{4.375}{0.375} = 7.32$$

With a level of significance at 0.05 and the df = 8 − 1 = 7, we find the critical t value to be 1.895. Note that the number of degrees of freedom is the number of pairs minus one, which is different than independent samples ($N_1 + N_2 - 2$). We obtain a t value of 7.32, which is greater than the critical value. We conclude that the diastolic blood pressure is significantly reduced after 3 months of jogging.

The paired subjects design is appropriate only when there is no carryover effect from one test to the next. When this effect occurs, it is difficult to determine whether the change in the posttest may be due to the treatment or the pretest. In general, the *t*-test is widely used in a variety of research situations examining mean difference. It is relatively simple compared to other statistical procedures. The numerator of the *t*-test is the observed difference between two groups; the denominator is an estimate of how much these two groups would be expected to be different by chance alone. The denominator is based on two factors: the sample size — the larger the sample, the less random differences to be expected — and the variability within the groups — the greater the variation, the greater the random differences to be expected.

When we form the alternative hypothesis, we may state that either two populations are different or one population mean is greater than the other. In the former case we want to know if the two means differ, regardless of which one is greater than the other. These statistics tests are called *two-tailed* tests.

Often, however, we are interested in finding out if one population mean (μ_1) is greater than the other (μ_2), but not vice versa. These statistical tests are called *one-tailed* tests. If it happens that $\mu_2 > \mu_1$ and the difference exceeds the critical value, we will conclude that the results are *not* significant.

At each level of significance, critical values are different for one- and two-tailed tests. The following chart gives critical values of Z for both one- and two-tailed tests at 0.10, 0.05, and 0.01 levels of significance:

	Level of significance		
	0.10	**0.05**	**0.01**
One-tailed critical values of Z	−1.28 or 1.28	−1.645 or 1.645	−2.33 or 2.33
Two-tailed critical values of Z	−1.645 or 1.645	−1.96 or 1.96	−2.58 or 2.58

Two-tailed tests are less powerful that one-tailed tests. For example, with ten degrees of freedom the *t* value greater than 2.76 is significant at the 0.01 level for a one-tailed test, whereas the *t* value for the two-tailed tests must reach 3.7 to achieve significance at the 0.01 level.

e. Analysis of Variance

Analysis of variance — often ANOVA for short — is a statistical technique for identifying sources of variation within the data. Analysis of variance is frequently used in statistics to analyze data obtained from experiments. Although analysis of variance can be considered an extension of the *t*-test, it is more widely applicable than the *t*-test. While the *t*-test can be used only to test two group means, analysis of variance can test two or more than two group means. Analysis of variance must have a dependent variable measured on an interval or ratio scale. The independent variable(s) are considered categorical, which means either an ordinal or a nominal scale. Independent variables are called factors in the analysis of variance. If the analysis of variance involves a single factor it is termed a one-way analysis of variance; if there are n factors involved in the analysis of variance it is referred to as an n-way analysis of variance.

As discussed before in Section II.F.2.d, we apply the *t*-distribution to evaluate the *t* statistic. In analysis of variance we use an F-distribution. We draw a sample from one population and draw a second sample from another population; then we calculate the F-ratio of two variances by the formula

$$F = \frac{s_1^2}{s_2^2}$$

Suppose we repeat this process an infinite number of times. The F-ratio would be obtained for each pair of samples. A frequency distribution can be established for these F-ratios. The

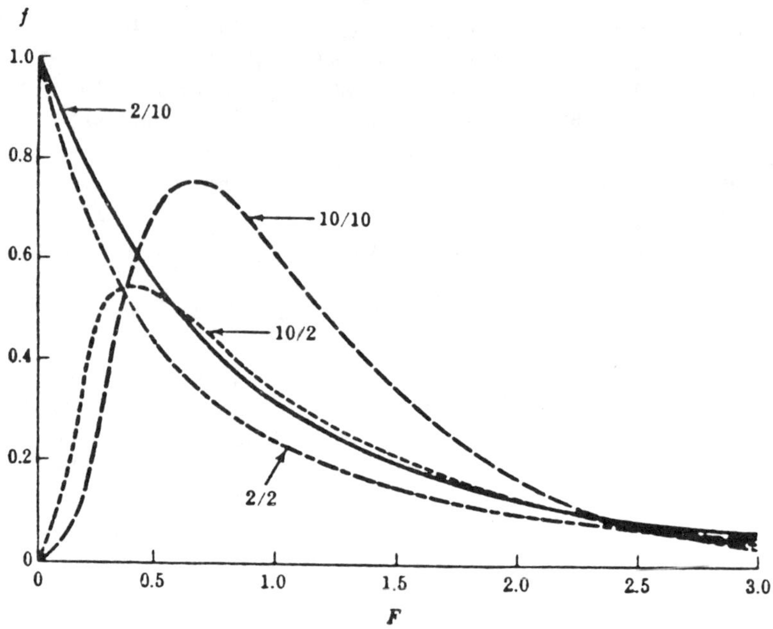

FIGURE 9. The F-distributions for various samples.

sampling distribution of these ratios is known as the F-distribution. The F-distribution is positively skewed ranging from zero to infinity. The F-distribution is a function of sample size (see Figure 9). Those larger sample sizes (i.e., larger degrees of freedom) tend toward normal distribution. Table 6 in the Appendix presents critical values of F-ratios with different degrees of freedom with 0.05 and 0.01 levels of significance.

There are a variety of research designs that utilize analysis of variance. We will introduce a few common designs and go through the computation procedures. The reader should generalize these models to similar analysis of variance situations.

f. One-Way Analysis of Variance — Completely Randomized Groups

i. Design

In a one-way analysis of variance there is only one independent variable with two or more levels of treatment. For example, an investigator may want to study the efficiency of three drugs in preventing human motion sickness induced in a laboratory situation. In this case the drug is the independent variable with three levels (e.g., three types of drugs). A group of subjects are randomly selected and assigned to one of the three drug conditions. The one-way analysis of variance permits the investigator to assess whether there is a significance difference among drugs in preventing motion sickness.

In the one-way analysis of variance situation, we formulate the null hypothesis that the k treatment means are equal (i.e., from the same population).

- H_0: $\mu_1 = \mu_2 = \mu_3 = \ldots = \mu k$.
- H_1: at least two of the treatment means are different.

Analysis of variance will partition the total variance of all subjects into two sources: between-subject variance and within-subject variance. The test statistic (the F-ratio) is the ratio between two variance estimates: between-subject variance is based on the dispersion of group means, while within-subject variance is based on the dispersion of scores within groups.

TABLE 9
The Length of Time (in Seconds) Obtained in Three Random Samples after Three Different Medicine Treatments

Group A		Group B		Group C	
x	x^2	x	x^2	x	x^2
12	144	13	169	19	361
16	256	15	225	23	529
10	100	9	81	20	400
9	81	11	121	16	256
13	169	14	196	15	225
8	64	10	100	11	121
10	100	12	144	14	196
7	49	7	49	8	64
9	81	11	121	12	144
6	36	8	64	12	144
100	1080	110	1271	150	2440

$$F = \frac{\text{Between-subject variance estimate}}{\text{Within-subject variance estimate}}$$

The F-ratio is likely to be large when between-subject variance is greater than within subject variance. The larger the F-ratio the greater the chance that the null hypothesis will be rejected. The F table should be consulted to determine if the probability of the observed F-ratio is greater than the level of significance. The following is an illustration.

A medical researcher may wish to compare the length of time for a group of subjects to respond to treatment by three different drugs, A, B, and C. A total of 30 subjects are randomly selected and assigned to three equal groups. Ten subjects in the first group are treated with medicine A, ten subjects in the second group with medicine B, and ten subjects in the third group with medicine C. The data are presented in Table 9.

The first step in analysis of variance is to compute the sum of squares between subject (SSb) and sum of squares within subject (SSw).

The formula for computing the between-subject variance is as the follows:

$$SSb = \frac{(\sum X_1)^2}{n_1} + \frac{(\sum X_2)^2}{n_2} + \ldots - \frac{(\sum X)^2}{N}$$

The value in the above example is

$$SSb = 100^2/10 + 110^2/10 + 150^2/10 - 360^2/30 = 140$$

The formula for computing the within-subject variance is as follows:

$$SSw = \sum X_1^2 - \frac{(\sum X_1)^2}{n_1} + \sum X_2^2 - \frac{(\sum X_2)^2}{n_2} + \ldots$$

The value in the above example is

$$SSw = 1080 - 100^2/10 + 1271 - 110^2/10 + 2440 - 150^2/10$$
$$= 331$$

TABLE 10
Summary of the Analysis of Variance of the Three Groups

Source of variance	SS	df	MS	F	*p*
Between subject	140	2	70	5.71	0.01
Within subject	331	27	12.26		
Total	471	29			

These sums of squares need to be converted into variance estimates by dividing each by its respective degrees of freedom. The variance estimates are also referred to as the mean square between (MSb) and the mean square within (MSw). The F-ratio is computed by dividing the mean square between subject by the mean square within subject.

There are two types of degrees of freedom associated with the F-test. The degrees of freedom for between subject is the number of groups minus one. We obtain this value as 3 − 1 = 2. The degrees of freedom for within subject is the sum of number of subjects in each group minus one. We obtain this value as 10 − 1 + 10 − 1 + 10 − 1 = 27.

Now the F-ratio is computed as

$$F = \frac{SSb/df}{SSw/df} = \frac{MSb}{MSw}$$

In the above example the computed F-ratio is

$$F = \frac{140/2}{331/27} = \frac{70}{12.26} = 5.71$$

We consult the F critical value ($\alpha = 0.05$) in Table 6 of the Appendix. We first locate the between-subject degrees of freedom (i.e., 2) in the first row and then locate the within-subject degrees of freedom (i.e., 27) in the first column. The entry value, 3.35, is the critical value at the 0.05 level of significance. Our obtained F-value, 5.71, is greater than the critical value, 3.35, so the null hypothesis is rejected. Table 10 summarizes of analysis of variance results.

The assumptions underlying the one-way analysis of variance are essentially the same as those of the *t*-test for two independent samples. When the F_{max} test is applied to only the two groups situation the group variance that is larger will be the numerator and the smaller one will be the denominator. The K degrees of freedom (see first row in Table 5 of the Appendix) will always be the number of groups (i.e., 2), and the v degrees of freedom (see first column in Table 5 of the Appendix) is the average of the two group sample sizes minus one. When the F_{max} test is used, with more than two groups in this case, the numerator is the largest variance of all groups and the denominator is the smallest variance of all groups. The K degrees of freedom is the number of groups (three in the above example), and the v degrees of freedom is the average of the sample sizes from all groups minus one [i.e., (10 + 10 + 10)/3 − 1 = 9]. Readers should not be confused by the degrees of freedom for F_{max} and the degrees of freedom for the F-test.

ii. Multiple Comparisons

When the overall F-test of group differences (more than two groups) is significant, this significance only suggests that at least two group means are different. The overall F-test does not tell which groups differ. Without knowing this information we may not be able

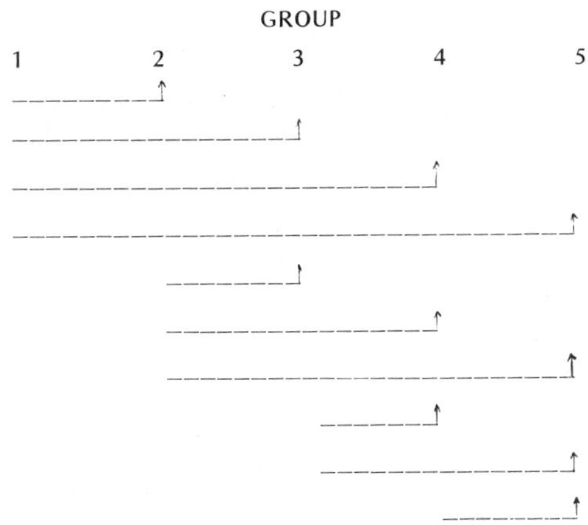

FIGURE 10. Schematic demonstration of possible pairs of comparisons for five groups.

draw appropriate conclusions about the data. A follow-up statistic called multiple comparisons helps identify which group means differ. If a factor has only two groups, no multiple comparisons are necessary because the overall F of the analysis of variance tests these *two* groups' means.

The key problem of multiple comparisons is to control Type I error. Readers are reminded that the Type I error occurs when two means are equal but the researcher declares them unequal. Suppose the overall F-test of analysis of variance of the treatment effect with five groups achieves significance at the 0.05 level. We wish to find out which groups differ. We may choose to compare all possible pair means in the absence of ideas about which groups differ. There are ten such comparisons altogether (see Figure 10).

It is noted that the number of possible comparisons can be obtained by

$$\frac{N!}{(N - k)!\, k!}$$

where N is the number of groups and k is the number of combinations (i.e., for pairwise comparisons k = 2).

When we compare the *t*-value of each such comparison with the $\alpha = 0.05$, the probability of making a Type I error at the 0.05 level is much greater than 0.05. The level of significance (say 0.05) applies only to that particular comparison and not to multiple comparisons. In multiple comparisons the joint probability of making a Type I error becomes

$$1 - (1 - \alpha)^k$$

where k is the number of comparisons.

With this formula, the joint probability of a Type I error for ten comparisons becomes

$$1 - (1 - 0.05)^{10} = 1 - 0.60 = 0.40$$

In other words, when we make ten comparisons with $\alpha = 0.05$, the probability of making at least one Type I error is 0.40. As the number of comparisons increases, the probability of making at least one Type I error also increases. A number of multiple comparison methods have been developed to control the Type I error rate; this will be discussed next.

g. A Priori and A Posteriori Comparisons

When a researcher plans to make multiple comparisons based on research interest rather than all possible pair means, it is called planned comparisons or *a priori* comparisons. For example, suppose five groups in the above example represent five different drugs. The first drug is the standard type. We wish to know if any of the other four new drugs is more effective than the standard one. All we need is the four multiple comparisons: each new drug compared with the standard drug. The corrected level of significance for multiple comparisons can be computed as α/n where n is the number of comparisons.

When there are four comparisons of interest the corrected is 0.05/4 = 0.0125. This means that with $\alpha = 0.05$ we actually have to use $\alpha = 0.0125$ for four multiple comparisons. This corrected α may be a more realistic error rate than all possible mean comparisons, which would yield an error rate at 0.05/10 = 0.005. We lose statistical power when we increase the number of comparisons. In *a priori* comparisons it is common practice to perform multiple comparisons (i.e., using the *t*-test) in lieu of the overall analysis of variance F-test, whether it is significant or not.

A posteriori comparisons, also called post hoc tests, are performed only after a significant global result has been found. *A posteriori* comparisons then test all possible pairs of group means. Various *a posteriori* comparisons have been developed, and the most widely used are least significant difference (LSD), Tukey honestly significant difference, and Scheffé's test. They differ in their way of controlling the error rate. LSD is the most powerful and Scheffé's test is the most conservative. In other words, the significant group difference found in the LSD test may not be significant in Scheffé's test. Extensive discussion on the various multiple comparisons methods can be found in books by Kirk[11] and Winer.[20]

i. Tukey's Method

Tukey's multiple comparison method is illustrated in the following steps:

Step 1 — Compute all possible comparisons between sample means. For our one-way analysis of variance example we will compute $X_1 - X_2$, $X_1 - X_3$, and $X_2 - X_3$:

$$X_1 - X_2 = 10 - 11 = -1$$

$$X_1 - X_3 = 10 - 15 = -5$$

$$X_2 - X_3 = 11 - 15 = -4$$

Step 2 — Divide each mean difference by the square root of MSw/n. The MSw is the mean square within subject and n is the number of subjects in any one group. Our MSw = 12.26 and n = 10; 12.26/10 = 1.1.

$$-1 / 1.1 = -0.91$$

$$-5 / 1.1 = -4.55$$

$$-4 / 1.1 = -3.64$$

Step 3 — Compare the Studentized Range Distribution table (see Table 7 in the Appendix). The first row of Table 7 indicates the number of comparisons, and the first column indicates the within-subject degrees of freedom. The critical value with 3,27 degrees of freedom at the 0.05 level of significance is 3.5. Any absolute difference between means divided by MSw/n that exceeds 3.5 is significant. In this case, group 3 differs from both group 1 and group 2 because 4.55 and $3.64 > 3.5$, while group 1 is not different from group 2 because $0.91 < 3.5$.

ii. Scheffé's Method

Scheffé's multiple comparison method is computed by the following formula:

$$F = \frac{(\overline{X}_1 - \overline{X}_2)^2}{MSw\ (1/n_1 + 1/n_2)\ (k - 1)} \text{ with df} = k - 1, N - k$$

where $\overline{X}$ is the group mean, MSw is the mean square within subject, n is the sample size, N is the overall sample size, and k is the number of groups.

Using the same three group comparisons with Scheffé's method we perform all possible group comparisons. Comparing the means for group 1 and group 2 we obtain:

$$F = \frac{(10 - 11)^2}{12.26\ (1/10 + 1/10)\ (3 - 1)} = \frac{1}{4.90} = 0.2$$

Comparing the means for group 1 and group 3 we obtain:

$$F = \frac{(10 - 15)^2}{12.26\ (1/10 + 1/10)\ (3 - 1)} = \frac{25}{4.90} = 5.1$$

Comparing the means for group 2 and group 3 we obtain:

$$F = \frac{(11 - 15)^2}{12.26\ (1/10 + 1/10)\ (3 - 1)} = \frac{16}{4.90} = 3.27$$

The critical value with degrees of freedom 2,27 at $\alpha = 0.05$ is 3.35 (see Table 6 in the Appendix), so we make the conclusion that group 3 is significantly different from group 1 ($5.1 > 3.35$). No other significant group difference has been found with Scheffé's criterion.

Remember that Tukey's method is only approximate for unequal group sizes while Scheffé's method is exact, even for unequal group sizes. Scheffé's method is not limited to two group means comparison. It is appropriate for examining all possible linear combinations, such as comparing the mean of group 3 with the mean of groups 1 and 2 combined.

h. Two-Way Analysis of Variance

In the above example there is only one experimental variable or factor with three experimental units (three groups with three different drugs); the analysis of variance for this situation is often referred to as one-way or simple analysis of variance. Very often we wish to examine the effects of several factors. A study with the combined effects of two or more independent variables is said to have a factorial design. There are many forms of factorial design, such as 2 × 2, 2 × 3, 3 × 3, 2 × 2 × 3 design, etc. Each number represents the levels of a factor. A 2 × 2 design means that there are two factors in the model and each factor has two levels. We will take a look at one of the many designs, 2 × 3, to illustrate the concepts and analysis.

Suppose we expand the design we just discussed about the length of response to three types of drugs. We also consider the way the drug is taken as one factor (e.g., fast intake and slow intake). In this case we will have two factors, one factor having two levels and another, three levels. All together there are six treatment combinations. We still select 30 subjects and assign them equally to one of the six treatment conditions. The design is presented schematically in Figure 11.

As before, we can also analyze the total variance into between-subject and within-subject variances. The difference is that the between-subject variance can be further subdivided into three component sources of variation. In other words, we can examine three effects:

	Drug Type		
Intake Type	A1	A2	A3
B1 (fast)	A1B1	A2B1	A3B1
B2 (slow)	A1B2	A2B2	A3B2

FIGURE 11. Schematic diagram of a 2 × 3 factorial design with five subjects in each cell.

TABLE 11
The Length of Time (in Seconds) Obtained in Three Random Samples after Three Different Medicine Treatments

	Group A1		Group A2		Group A3		
	X	X^2	X	X^2	X	X^2	
B1	12	144	13	169	19	361	
	16	256	15	225	23	529	
	10	100	9	81	20	400	
	9	81	11	121	16	256	
	13	169	14	196	15	225	
Sum	60	750	62	792	93	1771	Σ X = 215
B2	8	64	10	100	11	121	
	10	100	12	144	14	196	
	7	49	7	49	8	64	
	9	81	11	121	12	144	
	6	36	8	64	12	144	
Sum	40	330	48	479	57	617	Σ X = 145
	100	1080	110	1271	150	2388	

Grand X = 4739

(1) whether there is a significant difference between the length of time under a fast intake condition and under a slow intake condition; (2) whether there is a significant difference in time among the three types of medicine; and (3) whether or not the two factors, medicine type and intake type, have a combined effect on the length of time. To distinguish these effects we call the effects of the first and second analyses the main effects, whereas the third is referred to as the interaction effect. Data are presented in Table 11.

To compute the sum of squares total the following equation is used:

$$SSt = \sum X^2 - \frac{(\sum X)^2}{N}$$

TABLE 12
Summary of the Two-Way Analysis of Variance of the Three Groups

Source of variance	SS	df	MS	F	*p*
Between subject					
A	140	2	70	11.93	0.01
B	163.4	1	163.4	27.84	0.01
A × B	25.8	2	12.9	2.20	0.13
Within subject	140.8	24	5.87		
Total	470	29			

$$SSt = 12^2 + 18^2 + \ldots + 12^2 + 12^2 - \frac{360^2}{30}$$

$$SSt = 4791 - \frac{360}{30} = 470$$

To compute the sum of squares between subjects the following equation is used:

$$SSb = 60^2/5 + 40^2/5 + 62^2/5 + 48^2/5 + 93^2/5 + 57^2/5 - 360^2/30$$

$$= 329.2$$

Between-subject sums of squares are then broken down into three separate sums of squares: (1) the sum of squares between factor A, (2) the sum of squares between B, and (3) the sum of squares of interaction between factors A and B.

1. To compute the sum of squares for factor A:

$$SSa = 100^2/10 + 110^2/10 + 150^2/10 - 360^2/30 = 140$$

2. To compute the sum of squares for factor B (interaction):

$$SSb = 215^2/15 + 145^2/15 - 360^2/30 = 163.4$$

3. To compute the sum of squares for factors A and B:

$$SSab = 329.2 - 140 - 163.4 = 25.8$$

To compute the sum of squares within subject the following equation is used:

$$SSw = 470 - 329.2 = 140.8$$

Table 12 presents the summary results of analysis of variance.

When we examine the effect of two or more factors we must be aware of the possible interaction effect. Interaction effect means that the effect of one factor on the dependent variable is not the same with different levels of a second factor. This is better illustrated graphically. Figure 12 presents some hypothetical results from our drug type and intake type study.

In Figure 12(a) the effects of fast intake and slow intake (Factor B) show a parallel

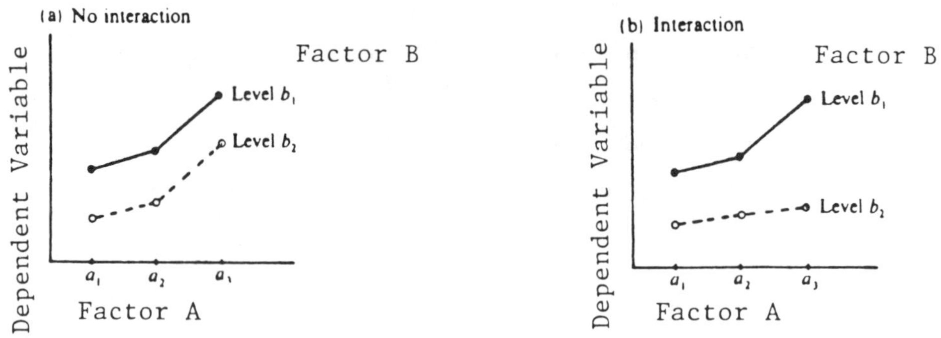

FIGURE 12. Illustrations of interactions. (a) No interaction; (b) interaction.

shape across the three drugs. In other words, the magnitude of difference between fast intake and slow intake is the same regardless of drug type. Also, the magnitude of difference among the three drug types is the same regardless of whether intake is fast or slow.

Figure 12(b) shows that the shapes of fast and slow intake are not parallel along the three types of drug. In other words, the effect of fast and slow intake depends on the type of drug. In this case we say that there is an interaction between the type of intake and the type of drug. In the factorial analysis of variance it is important to detect interaction effects first. If we were to analyze the difference between fast and slow intake effect, we would likely find that the difference would be smaller than it really is. This is because the effect of fast and slow intake types is based on the dependent variables averaged over three types of drug. If the interaction is overlooked and we conclude that there is no significant difference between fast and slow intake, we would certain mislead the readers!

It becomes obvious that the first test of analysis of variance in a factorial design is to examine the effect of interaction. When there is no interaction effect, we can then examine the main factor. However, when there is a significant interaction effect, the main factor should be ignored. Instead, we examine the difference of one factor at each level of another factor. In this example we can examine the fast and slow intake type for each drug type, rather than all drug types together. This test is sometimes called the test of simple effect.

i. Analysis of Variance with Repeated Measures

In between-subjects designs, subjects are given only one treatment condition. Often we want to assign a subject to each treatment condition under study. Designs with subjects repeating more than one treatment condition are called repeated measures designs. The two-related-samples design (i.e., in the paired *t*-test situation) is the simplest repeated measures design.

There are a number of advantages of using the repeated measures design. First, the repeated measures design increases the precision of the experiment. When the between-subject design is used there will be unavoidable subject variation from treatment to treatment no matter how they are randomly assigned to different treatments. The subject variation is excluded in the repeated measures design since the same subjects are assigned to each level of treatment. The repeated measures design is powerful because the reduced error variance is associated with greater sensitivity in detecting small differences in treatment effects. In contrast, the between-subject design uses different groups of subjects for each treatment condition, and the individual differences can result in a considerable amount of error variance, causing the design to be less sensitive to small differences in treatment effects.

Second, the repeated measures design can economize on the number of subjects. The one-way analysis of variance example given above requires 30 subjects in the experiment in order to have 10 on each drug type. However, using the repeated measures design of the same study, all together only ten subjects will be needed because each subject will receive all three types of drug rather than one subject receiving only one type of drug. Under certain circumstances, when large numbers of subjects are difficult to get, the repeated measures design is particularly helpful. The repeated measure design may also save time and money in experiments that require a pretrial and/or pretraining. When the same subjects will repeat different treatment conditions the pretrial or pretraining may only be necessary for the first treatment condition. However, in the between subjects design the subjects in each treatment have to receive a pretrial or pretraining.

Third, the repeated measures design is desirable if the researcher is interested in observing the treatment effect over a time period, such as a learning curve. However, if this is not what the researcher is interested in, it could confound the treatment effect because the effect could be caused by the preceding treatment. The effects the earlier treatments have on performance in the latter treatment is often called a carryover effect. A carryover effect can be desirable or undesirable, depending on the purpose of the study. For example, a specialist in sports injury is interested in finding out how three different kinds of knee braces may protect knee injuries. A number of subjects are selected randomly and each is given knee brace conditions. Knee movement is measured in each condition. The test for the first condition may affect the second and third. The greater movement of the third brace test may be due to the fact that the subject may become more flexible after a number of trials in the first and/or the second brace conditions. This carryover effect is obviously undesirable. The solution to this problem is to randomize the order of the treatment conditions.

There are also disadvantages in the repeated measures design. The repeated measures design requires rather restrictive assumptions — that is, the assumption of equality of variances and the assumption of equality of covariances. The equality of variances refers to the equal variability of scores in each of the treatment conditions. The equality of covariances refers to the consistency of measures across different treatment conditions. The assumption of equal covariances specifies that subjects should maintain their relative standing in the different treatment conditions. For example, if subjects score high in one treatment they should score relatively high in other treatments. If this condition is true there should be no significant treatment by subject interaction, which means that subjects respond in different ways to different treatment conditions. When these two assumptions are violated the probability of make a Type I error increases. However, the violation of these assumptions frequently occurs in behavioral studies.[20] The statistical tests of these assumptions and the necessary solutions when the assumptions are not satisfied are beyond the scope of this book. The reader should review References 7, 16, and 20.

j. One-Way Analysis of Variance with the Repeated Measures

Let us use a simple example to illustrate the one-way repeated measures analysis. We wish to examine which of three types of medicine will produce the desired effect in the most efficient length of time. Ten subjects are selected randomly. The response time is measured for each subject when given each medicine. The order in which the subjects receive the different types of medicine is random. Their measures are presented in Table 13.

The hypotheses for this problem are:

- H_0: $\mu_1 = \mu_2 = \mu_3 = \mu_4$.
- H_1: at least two of the treatment means are different.

The computations are as follows:

1. Compute the sum of squares for the treatment:

$$SSt = \frac{(\sum X_{t1})^2 + (\sum X_{t2})^2 + \ldots + (\sum X_{tn})^2}{n} - \frac{(\sum X)^2}{N}$$

TABLE 13
The Response Time for the Four Medicine Treatments

	Treatment (medicine type)				
Subject	**1**	**2**	**3**	**4**	**Total**
1	2	3	4	5	14
2	1	2	3	4	10
3	5	2	7	8	22
4	5	3	7	8	23
Total	13	10	21	25	69

where X_t is the sum of scores in each treatment condition, X is the sum of scores for all subjects in all treatment conditions, n is the number of subjects, and N is the number of subjects times the number of treatments.

$$SS_t = \frac{13^2 + 10^2 + 21^2 + 25^2}{4} - \frac{69^2}{16}$$

$$= 333.8 - 297.6 = 36.2$$

2. Compute the sum of squares for subjects:

$$SSs = \frac{(\sum X_1)^2 + (\sum X_2)^2 + \ldots + (\sum X_n)^2}{t} - \frac{(\sum X)^2}{N}$$

where ΣX_n is the sum of scores for each subject and t is the number of treatments.

$$SSs = \frac{14^2 + 10^2 + 22^2 + 23^2}{4} - \frac{69^2}{16}$$

$$= 327.3 - 297.6 = 29.7$$

3. Compute the sum of squares for scores:

$$\sum X^2 = 1^2 + 2^2 + \ldots + 8^2 = 373$$

4. Compute the sum of squares for treatment by subject interaction:

$$SS_{t \times s} = 373 - 333.8 - 327.3 + 297.6 = 9.5$$

The summary of the analysis is presented in Table 14.

The F test is

$$F = \frac{\text{MS treatment}}{\text{MS treatment by subject}} \text{ with df } (t - 1), (t - 1)(n - 1)$$

The F test for the treatment effect is obtained by 12.07/1.06 = 11.39, which is greater than the critical value 6.99 with degrees of freedom 3,9 at the 0.01 level of significance

TABLE 14
The Summary of One-Way Analysis of Variance with the Repeated Measures Design

Source of variation	SS	df	MS	F	p
Treatment	36.2	3	12.07	11.39	0.01
Subject	29.7	3	9.9		
Treatment by subject	9.5		1.06		

(see Table 5 in the Appendix). We reject the null hypothesis that the four treatment means are from the same population and accept the alternative hypothesis that at least two treatment means are from different populations. As we discussed before, the overall significant F does not tell us which two means differ. To determine which means differ we may perform multiple comparisons.

k. The Mixed Design — Designs with Both Between and Within Subject Factors

Researchers more often encounter a mixed design — one or more between-subject factors and one or more within-subject factors — than purely one design or the other. The schematic layout of a mixed design is presented below. In this design, between-subject factor A has two levels and within-subject factor B has four levels. This is often called a 2 × 4 (A × B) design with the repeated measures on the B factor. The factorial combination of these two factors leads to eight cells: A1B1, A2B2,..., A2B4. Keep in mind that a between-subject factor is one in which *independent subjects* are given *different levels* of the factor. A between-subject factor is usually specified with a grouping variable. A within-subject factor is one for which the *same subject* is measured on *each level* of the factor. This is also called a repeated measures factor.

Schematic Illustration of a 2 × 4 Mixed Design

	Within-subject factor B			
Between-subject factor A	**B1**	**B2**	**B3**	**B4**
A1	A1B1	A1B2	A1B3	A1B4
A2	A2B1	A2B2	A2B3	A2B4

The data analysis involves three tests:

1. Is there a significant difference between two levels of factor A? This is to test the main effect of A.
2. Is there an overall significant difference among the four levels of factor B? This is to test the main effect of B.
3. Is there an interaction between two factors, A and B?

As we described previously, in most factorial designs the interaction should be examined first.

Let us use an example to illustrate the computation. Data from a 2 × 2 (A × B) design with repeated measures on factor B are presented in Table 15.

In this 2 × 2 with one repeated factor design, we need to compute the following five sums of squares:

TABLE 15
Data from 2 × 2 Mixed Design with Repeated Measures on Factor B

	B1	B2	Sum
A1	2	7	9
	1	4	5
	3	3	6
Sum	6	14	20
A2	3	8	11
	2	5	7
	6	8	14
Sum	11	21	32
	17	35	52

1. Compute sum of squares of A effect:

$$\text{SS a} = \frac{20^2 + 32^2}{6} - \frac{52^2}{12}$$

$$= 237.3 - 225.3 = 12$$

2. Compute sum of squares of B effect:

$$\text{SS b} = \frac{17^2 + 35^2}{6} - \frac{52^2}{12}$$

$$= 252.3 - 225.3 = 27$$

3. Compute sum of squares of A × B interaction:

$$\text{SS a} \times \text{b} = \frac{6^2 + 14^2 + 11^2 + 21^2}{3} - 252.3 - 237.3 + 225.3$$

$$= 264.7 - 252.3 - 237.3 + 225.3 = 0.4$$

4. Compute the sum of squares of subject within A:

$$= \frac{5^2 + 9^2 + 6^2 + 7^2 + 11^2 + 14^2}{2} - 237.3$$

$$= 254 - 237.3 = 16.7$$

5. Compute the sum of squares of B × subject within A:

$$290 - 254 - 264.7 + 237.3 = 8.6$$

Table 16 presents the summary results.

TABLE 16
Summary of the Analysis of Variance

Source	SS	df	MS	F	p
A	12	1	12	2.9	0.05
Subject/A	16.7	4	4.2		
B	27	1	27	12.5	0.01
A × B	0.4	1	0.4	0.15	
B × subject/A	8.67	4	2.17		

The reader is cautioned that the repeated measures design does not have a within-subject variance estimated that can be used for the denominator of the F-ratio test as in the between-subject design situation. Instead, the denominator for the A factor is the mean square of subjects within the A factor. The denominator for testing the B and A × B interaction is the mean square of B × subjects within A.

The F-ratio test for A factor is

$$F = 12/4.2 = 2.9$$

The F-ratio test for B factor is

$$F = 27/2.17 = 12.5$$

The F-ratio test for the A × B interaction is

$$F = 0.4/2.17 = 0.5$$

Compared to the F table, the value indicates that both A and B factors are significant at the 0.05 and 0.01 levels, respectively, while the A × B interaction is not significant.

G. REGRESSION ANALYSIS

Regression and correlation are related in terms of their analysis, but the purpose is different. Correlation analysis provides the investigator with the correlation coefficient that measures the strength of the relationship between two variables. Regression provides the researcher with the equation that can predict the criterion (dependent) variable with one or more predictors (independent) variables. In addition, regression analysis provides variance measures that allow researchers to assess the accuracy of the prediction equation. In the case of multiple predictors, the regression analysis will assess the relative importance of each predictor in its contribution to the predictor of the criterion variable. There are many instances in which regression analysis can be applied. For example, a researcher studied malnutrition, which might be affected by the number of children in a family, in a preschool child population. Using a house-to-house survey the researcher obtained the weight of a sample of preschool children to determine their degree of malnutrition. The effect of family size on malnutrition was examined by grouping the children according to the number of living children in their families. The regression technique helped to determine if malnutrition tended to increase with the number of children in the family. Another example: a study examined the relationship between subjects' intracranial pressure, obesity, and age. The multiple regression analysis quantitatively described whether subjects who were overweight and older would have higher blood pressure than those who weighed less and were younger.

To illustrate what a regression equation looks like, we will look at the values of subjects

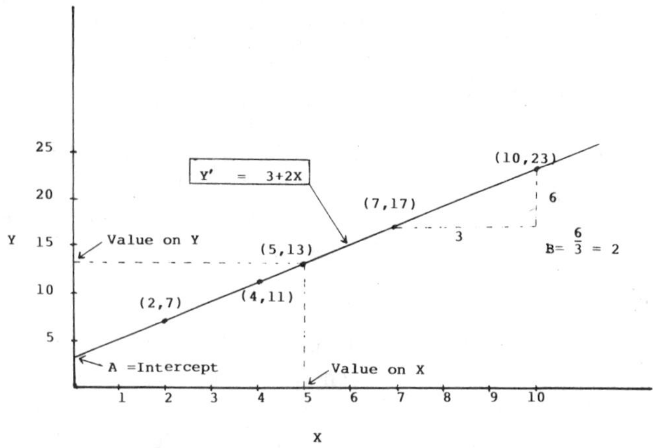

FIGURE 13. Characteristics of a simples linear equation.

on two variables to gain some insight into the nature of the regression equation. Five subjects' scores on variables X and Y are presented as follows:

	Score on variable X	**Score on variable Y**
Subject 1	2	7
Subject 2	5	13
Subject 3	4	11
Subject 4	10	23
Subject 5	7	17

Our question is whether any systematic relationship can be identified. The scores of the two variables demonstrate that the value in variable Y is greater than the value of X for the same subject. A closer examination shows that the score on variable Y is exactly twice the score of variable X, plus three. If we plot these two sets of scores in a two-dimensional scale with variable Y on the vertical line and variable X on the horizontal line, we will generate the graph shown in Figure 13. Clearly, there is a straight line relationship between the two variables.

If the relationship between these two variables is specified in a mathematical form, it will be

$$Y' = 3 + 2X$$

This equation is just a specific instance from the above two data sets. The general equation of a regression line is

$$Y' = A + BX$$

Y′ is the dependent variable or criterion variable from the population; X is the independent variable or predictor variable. This equation indicates that there is a linear regression of Y′ on X. The values of A and B are estimated from the data.

The value of B is called the slope of the regression. When predicting the Y′ score from this equation we have to multiple the score of X by the value of B. B is the rate of change in the score of Y′ for a unit change in the score of X. Generally, the higher the value of B, the stronger the relationship between X and Y′ and the steeper the slope. The value of A, called the intercept of the regression equation, represents the value of Y′ when $X = 0$. This value is generally not very interesting.

Regression can also take the form of nonlinear relationships between dependent and independent variables. In this chapter we will limit our discussion to only linear regression. The interested reader may consult the book by Neter and Wasserman.[13] The variables in the linear regression analysis, both dependent and independent variables, are interval or ratio variables.

Figure 13 shows that all five pairs of data lie exactly on the regression line. In reality, we can hardly find such a perfect relationship between two variables. Very often we find that the scores of the two variables are scattered. We try to find the "best-fit" line that passes through the cluster of data points. If we compute the vertical distance (deviation) between each data point and the best-fit line, then square each distance and sum them, such a value would be the smallest among all the possible regression lines [(i.e., $\Sigma(Y - Y')^2$) = smallest]. The most common approach to "best-fitting" a straight line to the data is the *least squares method.* The least squares method yields the smallest error estimation. The first step of regression analysis is to find the slope B and intercept A. Based on the least squares criterion, the slope B can be estimated by

$$B = \frac{\sum (X - \overline{X})(Y - \overline{Y})}{\sum (X - \overline{X})^2}$$

Another equivalent formula for the slope B is

$$B = \frac{N \sum XY - (\sum X)(\sum Y)}{N \sum X^2 - (\sum X)^2}$$

The value of the intercept, A, is computed by

$$A = \frac{(\sum Y)(\sum X^2) - (\sum X)(\sum XY)}{N \sum X^2 - (\sum X)^2}$$

or by

$$A = \overline{Y} - B\overline{X}$$

We will go through the numerical calculations involved in determining the values of B and A. Table 17 presents the hypothetical data of 12 healthy subjects. Variable Y is the diastolic blood pressure measured by the standard technique, while variable X is the diastolic blood pressure measured on a newly developed automatic blood pressure instrument. The study attempted to show that the new method is accurate (i.e., automated values approximate the standard value). Ten measurements were taken by both the old and the new methods, and the means were computed. Based on these data, we can evaluate the accuracy of the new technique, using the standard method as the criterion.

The following formula is applied for computing B:

$$B = \frac{N \sum XY - (\sum X)(\sum Y)}{N \sum X^2 - (\sum X)^2} = \frac{(12 * 54{,}107) - (798 * 813)}{(12 * 53{,}142) - (798)^2} = 0.57$$

TABLE 17
The Means of Diastolic Blood Pressure by Old and New Method and Calculations of Slope B and Intercept A

	X	Y	XY	X^2	Y^2
	65	68	4,420	4,225	4,624
	63	66	4,158	3,969	4,356
	67	68	4,556	4,489	4,624
	64	65	4,160	4,096	4,225
	68	69	4,692	4,624	4,761
	62	66	4,092	3,844	4,356
	68	70	4,760	4,624	4,900
	66	65	4,290	4,356	4,225
	68	71	4,828	4,624	5,041
	67	67	4,489	4,489	4,489
	69	68	4,692	4,761	4,624
	71	70	4,970	5,041	4,900
Sum	798	813	54,107	53,142	55,125

The following formula is applied for computing A:

$$A = \frac{(\sum Y)(\sum X^2) - (\sum X)(\sum XY)}{N \sum X^2 - (\sum X)^2}$$

$$= \frac{(813 * 53{,}142) - (798 * 54{,}107)}{(12 * 53{,}142) - (798)^2} = 30.1$$

The regression equation is

$$Y' = 30.1 + 0.57X$$

If we let X = 62 and Y′ = 65.4, and also let X = 72 and Y′ = 71.1, we can plot the regression line. This line is the best-fit line to describe the relationship between the sample values of the old method and the new method (see Figure 14).

With the regression equation we will be able to predict the true measure by the new method. For example, if one individual obtains a blood pressure of 65.6 from the new method, the predicted blood pressure from the standard method is

$$Y' = 30.1 + (0.57 * 65.5) = 67.4$$

Let us examine Figure 14 closely. None of the points actually fall on the best-fit line. In other words, the actual scores deviate from the predicted scores. It is obvious that the bigger the deviations, the poorer the prediction. We must evaluate the regression equation to determine whether or not it adequately describes the relationship between the two variables and to determine whether or not we can apply the equation effectively for prediction and estimation purposes. Two methods to do so, the coefficient of determination and the analysis of regression, will be discussed next.

One method of evaluating the regression equation is to compare the scatter of the scores about the regression line with the scatter about $\overline{Y}$, the mean of the sample values of Y. Figure 14 provides a visual inspection of the relative magnitudes of the scatter of the scores from $\overline{Y}$ and of the regression line. The line representing $\overline{Y}$ is shown as the horizontal line.

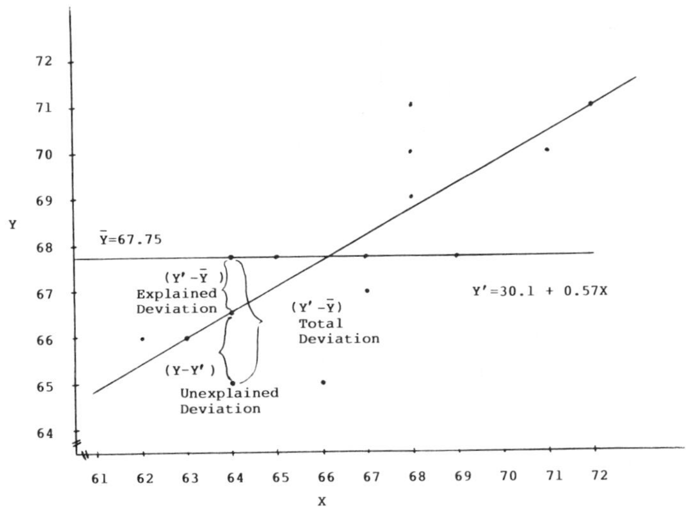

FIGURE 14. Scatter diagram showing deviation about Y and the regression line.

Regardless of the scores of X, $\overline{Y}$ remains constant. However, the deviation of scores about the regression line is much less than the deviation about the $\overline{Y}$ line. It seems apparent that the regression line provides a good fit for the data. The measure of R^2 provides an objective measure of how good the fit is. The following description will help the reader understand the source of the R^2 and what it really means.

There are three deviations to be considered in regression analysis. First, any deviation between the observed score and the mean of the measure is called the total deviation. This is represented by the vertical distance (see Figure 14) between any Y scores and the mean $\overline{Y}$ and is calculated by $(Y - \overline{Y})$. Second, any vertical distance between the regression line and the $\overline{Y}$ is called the "explained" deviation and is computed by $(Y' - \overline{Y})$. This value shows the amount of deviation reduced from the total deviation by fitting the regression line to the data. Third, the vertical distance between the observed score and the regression line is called the unexplained deviation or residual and is computed by $(Y - Y')$. This value represents the portion of the total deviation not "explained" by fitting the regression line.

The relationship among three deviations is expressed as

$$\text{Total deviation} = \text{explained deviation} + \text{unexplained deviation}$$

These three deviations are measured mathematically by the sum of the squared deviations, so the relationship is expressed as follows:

$$(Y - \overline{Y})^2 = (Y' - \overline{Y})^2 + (Y - Y')^2$$

R^2 is defined as follows:

$$R^2 = \frac{(Y' - \overline{Y})^2}{(Y - \overline{Y})^2} = \frac{\text{Explained sum of squares}}{\text{Total sum or squares}}$$

We can compute the total sum of squares (SS) by

$$\sum (Y - \overline{Y})^2$$

or equivalent formula

$$\sum Y^2 - (\sum Y)^2 / N$$

For the data in the above example, we obtain

$$\text{Total SS} = 55125 - (813)^2 / 12 = 44.3$$

We can next compute the explained sum of squares (SS) by

$$\sum (Y' - \overline{Y})^2$$

or the computational formula

$$B^2 \sum (X - \overline{X})^2$$

or

$$B^2 [\sum X^2 - (\sum X)^2 / N]$$

For the data in the above example we obtain

$$\text{Explained SS} = (0.57)^2 [53142 - (798)^2/12] = 24.4$$

The unexplained sum of squares (although we may not use it now) can be computed by

$$\text{Unexplained SS} = \text{total SS} - \text{explained SS} = 44.3 - 24.4 = 19.9$$

We then compute

$$R^2 = \frac{\text{Explained SS}}{\text{Total SS}} = \frac{24.4}{44.3} = 0.55$$

Thus, the regression of Y on X explains approximately 55% of the total variability in Y. In other words, the new blood pressure technique is 55% as accurate as the standard method.

R^2, also called the coefficient of determination, measures the proportion of the variance of the dependent variable that is accounted for by the independent variable. In other words, R^2 measures the closeness of fit of the regression equation to the same data. The better the fit of the regression line, the close R^2 will be to 1; the poorer the fit of the regression line, the closer R^2 will be to 0.

The analysis of regression will test the hypothesis that there is no relationship between the independent and the dependent variables in a regression equation. In the analysis of regression, the total sum of squares for the dependent variable is partitioned into a sum of squares for regression and a sum of squares for residuals. This is the same way we partition the total sum of squares in the analysis of variance. The sum of squares for regression is equivalent to the sum of squares between subjects, and the sum of squares for residuals corresponds to the sum of squares within subjects. It is possible, therefore, to test a null hypothesis of no linear regression between the dependent and the independent variable by analysis of variance.

The hypotheses are formed as follows:

- H_0: there is no linear regression between the dependent and the independent variable.
- H_1: there is a linear regression between the dependent and the independent variable.

TABLE 18
Summary of Analysis of Regression

Sources	SS	df	MS	F	*p*
Regression	24.4	1	24.4	12.2	0.01
Residual	19.9	10	2.0		
Total	44.3	11			

The summary results of the analysis of regression are presented in Table 18.

The degrees of freedom value for the regression is the number of independent variables — in this example, 1. The degrees of freedom value for the residual is N − K − 1, where K is the number of independent variables and N is the number of subjects. We obtain 12 − 1 − 1 = 10 for this example. The critical value for F(1,10) at the 0.01 level is 10.04. The computed F value of 12.2 is greater than 10.04; therefore, we reject the null hypothesis. We conclude that the data of this sample provide sufficient evidence to indicate that there is a linear regression between the independent and dependent variables.

1. Significance Test of Slope B

An alternative method of evaluating the sample regression equation is to test the regression slope, B. The significance test of B is of considerable interest. If B is not significant (i.e., B = 0), then there is no relationship between the dependent variable and the independent variable. The regression line is horizontal, and a change in the independent variable is not associated with a change in the dependent variable. The significance of B can be tested either by examining the confidence interval or, more conveniently, by evaluating the *t*-test. The test statistic for deciding whether or not B = 0 is

$$t = \frac{B}{s_B}$$

where s_B (square) is the standard error of B; it is defined as

$$s_B = \frac{\text{MS residual}}{\sum (X - \bar{X})^2} = \frac{\text{MS residual}}{\sum X^2 - (\sum X)^2/N}$$

The computed MS residual is 2.0, which can be found in Table 17. We compute

$$s_B = \frac{2.0}{53{,}142 - (798)^2/12} = 0.16$$

The values in the denominator are taken from the previous table. The test statistic is then

$$t = \frac{0.57}{0.16} = 3.56$$

With df = (N − 2) = 10 and the two-tailed level of significance of 0.01, the critical *t* value is 3.169. Since 3.56 > 3.169, we conclude that B is not 0 and that there is a linear relationship between the dependent and the independent variables.

B may be positive or negative. A positive B indicates that the dependent variable tends to increase as the independent variable increases. This is called a direct linear relationship between two variables. A negative B indicates that as the independent variable increases the dependent variable tends to decrease. This is called an inverse linear relationship between two variables.

2. Confidence Interval for B

We can also test the null hypothesis that B = 0 by means of a confidence interval for B. The confidence interval is given by

$$B \pm t(1 - \alpha/2; n - 2)s_B$$

where s_B is the standard error of B (0.16 as computed previously). t is the table value based on the level of significance and the degrees of freedom. For our example, constructing a 95% confidence and N = 12, the t will be $t(1 - 0.05/2; 12 - 2) = t(0.975,10) = 2.228$. Then the confidence interval for B is

$$0.57 \pm 2.228(0.16) \qquad 0.214, 0.926$$

This means if we repeatedly draw samples of 12 subjects from the population and compute a confidence interval for B as described above, 95% of these intervals (0.214 – 0.926) would include the parameter B. In other words, we are 95% confident that the single interval we have constructed includes the parameter B.

If the interval includes 0, the null hypothesis B = 0 is accepted; otherwise, the null hypothesis is rejected. Since our interval 0.214 to 0.926 does not include 0, we conclude that B is not 0 and that there is a linear relationship between the independent and dependent variables.

3. Assumptions Association with Simple Regression

There are several assumptions related to the simple regression analysis:

1. The values of the dependent variable (Y) are distributed normally. For each value of the independent variable (X) there is an associated normal distribution of the dependent variable Y.
2. The variances of the Y distributions are equal; this is called homoscedasticity.
3. The dependent variable Y values are statistically independent of each other. This means that the values of Y selected at one value of X do not depend on the values of Y selected at another value of X.
4. The means of the subpopulation of Y lie on the same straight line with any given value of X.

4. Multiple Linear Regression

Simple regression analysis can be extended to multiple linear regression involving two or more independent variables.

The general form of the multiple regression equation is:

$$Y' = A + B_1X_1 + B_2X_2 + \ldots + B_kX_k$$

where Y′ represents the predicted value of Y, A is the intercept, and the B_i are regression coefficients.

As in simple regression, the least squares method is used to determine A and B_i so as to yield the best multiple regression equation. This best-fit multiple regression equation will minimize the sum of squared deviations between the observed Y values and the predicted Y′ values $\Sigma(Y - Y')^2$.

After we have obtained the multiple regression equation, we must evaluate it to determine whether it adequately describes the relationship between the dependent variable and the independent variables. The evaluation of the equation includes the test of overall significance (i.e., the F test), the test of individual B coefficients, and multiple R^2.

5. Overall Significance F Test of the Equation

Regression analysis per se does not suggest a causal relationship between the dependent and independent variables. In this sense, regression may be considered descriptive statistics. However, researchers are often interested in generalizing the sample data to a population. This generalization refers to estimating population parameters from sample regression statistics or to testing statistical hypotheses about the population parameters.

The "overall" significance test for the regression equation uses statistical inference procedures to test the null hypothesis that the multiple correlation is zero. The test statistic employed for the overall significance test is

$$F = \frac{\text{SS regression} / k}{\text{SS residual} / (N - k - 1)}$$

or equivalent formula

$$F = \frac{R^2 / k}{(1 - R^2) / (N - k - 1)}$$

where SS regression is the sum of squares explained by the overall regression equation, SS residual is the sum of squares unexplained, k is the number of independent variables in the equation, and N is the sample size. Note that the formula also may be written in the squared multiple correlation coefficient.

The overall null hypothesis, H_0: R = 0, is equivalent to the null hypothesis that all regression coefficients in the population are equal to zero, i.e., H_0: $B_1 = B_2 \ldots B_k = 0$. The alternative hypothesis, H_1: R = 0, means that one or more regression coefficients are not zero. The overall significant F does not indicate which regression coefficient values are not zero; therefore, follow-up tests for each individual regression coefficient are necessary.

6. Follow-Up B Test

In the simple regression analysis, the test for B is simple because any proportion of the explained sum of squared deviations is attributable to this single independent variable. In a multiple regression situation, the explained sum of squares is attributed to all independent variables in the equation. Testing of B coefficients involves a breakdown of the explained sum of squares into components attributable to each independent variable in the equation. One mode of decomposition is called the standard regression method. In this method each variable is examined assuming all other variables had been included in the regression equation. In other words, only the increment portion of the explained sum of squares in R^2 due to the addition of a given variable is considered and tested. This can be done either by the *t*-test or the F-ratio test,

$$F = \frac{\text{SS increment due to X}}{\text{SS residual} / (N - k - 1)}$$

or equivalently

$$F = \frac{R^2\, y(i.12, \ldots, k)}{(1 - R^2)y.12, \ldots, k) / (N - k - 1)}$$

The R^2 y(i.12, . . . , k) refers to the R^2 change after a given variable, X_i, is added to the equation that contains all the other independent variables. If the F-ratio is greater than the critical value of F at the specified level of significance, we conclude that this variable (X_i) is significant and should be included in the regression equation; otherwise, the variable is excluded from the equation.

7. Multiple R^2

It is desirable to know the total explanatory power of a set of independent variables when combined. Multiple correlation (R) is the correlation between one dependent variable and more than one independent variable. The square of R (i.e., R^2) signifies the proportion of variance in the dependent variable explained by a set of independent variables. However, the R^2 is always regarded as consisting of a proportion of chance relationships; thus, it is biased upward. The degree of R^2 inflation is related to the size of the sample (N) and the number of variables (k). There is no absolute mathematical guidance for the number of subjects and the number of variables used in the regression model. The rule of thumb is that there should be at least ten times as many subjects as the number of variables. In other words, the subject-to-variable ratio should be at least 10:1. Any ratio below this may result in substantial R^2 inflation. Nunnally[15] gives a detailed description of this topic.

To estimate the unbiased R^2, subtract the amount of R^2 inflation from any computed R^2. This assessment can be done as follows:

$$\text{Unbiased } R^2 = R^2 - (k - 1) / (N - 1)$$

where k is the number of variables and N is the number of subjects.

For example, if we have obtained an R^2 of 0.61 with five variables on a sample of 20, then the unbiased R^2 is

$$\text{Unbiased } R^2 = 0.61 - (5 - 1) / (20 - 1) = 0.40$$

Assuming the sample is 50 (subject-to-variable ratio is 10:1) instead of 20, the unbiased R^2 becomes

$$\text{Unbiased } R^2 = 0.61 - (5 - 1) / (50 - 1) = 0.53$$

The example illustrates that a small sample size is associated with greater R^2 inflation. To reduce the amount of R^2 inflation, a greater subject-to-variable ratio than 10:1 should be considered.

8. Search for "Best" Set of Independent Variables

One of the most challenging jobs in multiple regression is to select a set of independent variables in the regression equation. We wish to identify a regression equation based on relatively few independent variables, yet which accounts for virtually all the variance that could be explained if the entire set of predictor variables was used.

When a large set of independent variables is compiled, a screening procedure may be necessary before regression analysis to exclude variables that (1) may not be fundamental to the research problem. Previous research, literature review, and theory will help to make the decision; (2) statistically may be subject to large measurement errors. Descriptive statistics such as standard deviation, variance, range, skewness, kurtosis, etc. will help to make this decision; (3) may have a low correlation with the dependent variable. The simple correlation of each independent variable and the dependent variable may be examined; and (4) may duplicate other independent variables in this set. A zero-order intercorrelation among all independent variables may be obtained. A duplicate may be the independent variable that is highly correlated with another independent variable.

Typically, after the initial screening the number of independent variables that remains may still be large. A number of selection methods for regression analysis to help find the best set of independent variables will be introduced. These selection methods involve extremely heavy computations. A statistical package (SAS, SPSSx, etc.) may be utilized.

Four selection methods will be introduced as follows:

1. Forward selection. The investigator sets the criterion for including the independent variables in the regression equation. The commonly used criterion is the level of significance (i.e., $p < 0.05$). Using $p < 0.05$ as the criterion, the significant independent variable with the lowest p value (most significant) will be selected to enter the regression equation. Then the significance of all the remaining variables will be tested, adjusting for the effect of the variable already entered. If the most significant variable meets the entry criterion, it is included in the equation. This process continues until no variable meets the criterion.
2. Backward elimination. Backward elimination is the opposite of forward selection. The procedure begins with all independent variables in the model and then drops the least significant variable having a $p > 0.05$ (assuming the eliminating criterion is $p > 0.05$). Then the significance of each variable in the equation is tested again. If the variable with the least significance (i.e., the highest p value) meets the elimination criterion, the variable is dropped. This process continues until no further independent variables can be dropped.
3. Stepwise solution. Stepwise solution, the combination of forward and backward selection, is probably the most widely used method. The first step of stepwise solution is to identify the most significant variable that meets the entry criterion (i.e., $p < 0.05$), as in the forward method. Then the significance of all the other variables is tested, adjusting the variables already in the equation, and the most significant variable that meets the entry criterion is included in the equation. The next step is to examine whether any of the independent variables already in the equation may be nonsignificant (due to the entry of the new independent variables) and, hence, should be dropped. This is achieved by testing the significance of each variable, adjusting the effect of the rest of the variables in the equation. If the p value of any variable is greater than the dropout criterion (i.e., $p = 0.10$), it is eliminated; otherwise, the variable is retained. This process continues, each time adding the most significant variable that meets the entry criterion while adjusting all variables already in the equation, and each time deleting the variable that no longer falls below the dropout criterion, after adjusting the rest of the variables. The process continues until no further independent variables can be either added or deleted.

   The advantage of the stepwise method is that researchers can identify a regression equation based on relatively few independent variables, but which accounts for a similar amount of variance when compared to the larger number of independent variables. The disadvantage of using the stepwise solution is that a researcher may not have any control over the theoretically important variables that may be excluded from the regression equation, especially when these variables are highly correlated with other independent variables.
4. All possible regression approach. This procedure calls for an examination of all possible regression equations involving the potential independent variables and a seach for the "best" equation according to a predetermined criterion (i.e., p or R^2). This approach starts all possible variable combinations one at a time, two at a time, three at a time, and so on. The advantage of this approach is that the researcher will be able to compare

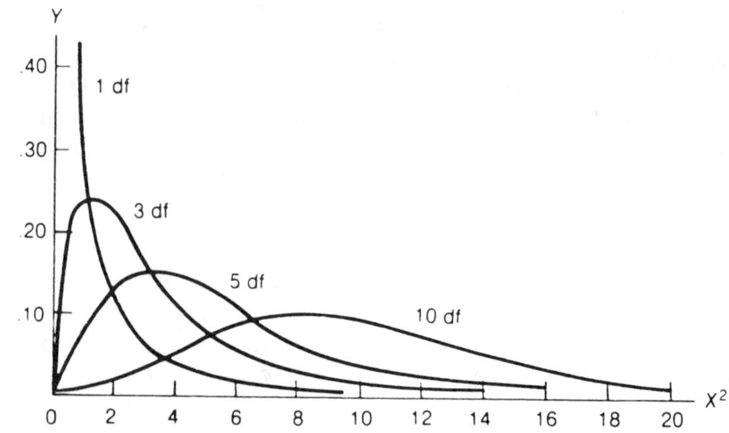

FIGURE 15. Chi-square distributions for one, three, five, and ten degrees of freedom.

the results of all possible combinations and select the best equation. However, this approach is extremely time-consuming and may lead to increased chance variance.

H. CHI-SQUARE ANALYSIS

The statistical techniques we have discussed in the previous sections require that the data for analyses consist of measurements — scores of some type of test. Sometimes the data obtained from a test consist of frequency counts rather than measurements, i.e., the number of individuals, objects, or events that fall into different categories of a variable. For example, consider the case in which some medical product is manufactured, under identical conditions, by two factories. A product manager suspects that one factory produces substantially more defective products than the other. The manager then randomly samples 500 products from each factory and counts the number of products failing to meet quality standards. Based on these frequency results, the manager may determine if the two factories are significantly different in manufacturing the medical product. This example demonstrates two characteristics: (1) the data consist of counts or frequencies of defective products from the sampled 500 products, and (2) there are expected frequencies of defective products which can be compared with the observed frequencies. Such a comparison can lead to some meaningful conclusions. If conditions are identical in the two factories, we may expect that approximately the same number of items would fail to meet the quality standards. The question is whether or not the discrepancies between observed and expected frequencies are large enough to reject the assumption on which the expected frequencies are based. The statistical technique commonly used to provide answers to such questions is called the chi-square statistic. The chi-square distribution is a sampling distribution, and it is skewed sharply to the right for small samples. As N increases, however, the distribution becomes more symmetrical and normally distributed (see Figure 15). We will not discuss the mathematical properties of the chi-square distribution. Interested readers should read the book by Lancaster.[12]

The chi-square test is as follows:

$$\chi^2 = \frac{\sum (O - E)^2}{E}$$

where O is the observed frequency and E is the expected frequency.

1. Chi-Square Tests of Goodness of Fit

The chi-square test of goodness of fit is used to determine whether an observed frequency distribution differs significantly from a hypothesized (expected) frequency distribution.

Suppose a medical company is interested in the number of sales of a newly produced drug. The company determined the number of drugs sold during the first 3 months after the drug was marketed. The number of cases sold obtained each month is as follows:

	Month			
	1	**2**	**3**	**Total**
Number sold	40	50	60	150

We want to know if there is a monthly increase in the number of sales. If there is no change in numbers, we would expect that the same number of sales would occur each month. In other words, it is expected that the total number of sales would be distributed uniformly over the 3 months. Following this line of reasoning, the hypotheses may be formulated:

- H_0: the number of drug sales follows a uniform distribution.
- H_1: the number of drug sales is not distributed uniformly.

If the data are distributed uniformly, the expected frequency would be as follows:

	Month			
	1	**2**	**3**	**Total**
Number sold	50	50	50	150

Assuming there is no change during the 3-month period, the expected number of sales is 150/3 = 50. In other words, the expected frequency is 50. To find out whether or not the observed responses are significantly different from expected responses, a chi-square test is performed:

$$\chi^2 = \frac{(40 - 50)^2}{50} + \frac{(50 - 50)^2}{50} + \frac{(60 - 50)^2}{50} = 4.0$$

The degrees of freedom are computed by the number of categories minus one; $R - 1 = 3 - 1 = 2$. Referring to the chi-square Table 8 in the Appendix, we see that with two degrees of freedom a chi-square value of 5.99 or greater is the critical value ($\alpha = 0.05$). Our computed chi-square value is smaller than 5.99; we accept the null hypothesis that the number of drug sales follows a uniform distribution. This means that the observed differences over 3 months could have occurred by chance alone.

2. Chi-Square Tests of Independence

The application of the chi-square test is not limited to a single dimension. By extending the above one-way classification table, we can arrive at a two-way classification table in which the observed frequencies occupy i rows and j columns. Such a table is called a contingency table.

The two-way classification table is used more frequently than the one-way classification table. Chi-square is useful in determining whether two nominal or ordinal measures are associated or independent. This type of chi-square is called chi-square test of independence. Two measures of classification are said to be independent if the frequency distribution of one criterion does not depend on the frequency distribution of the other criterion. If the

TABLE 19
The Observed Frequencies of Number of Sales During a 3-Month Period for Two Different Strategies

Month	1	2	3	Total
Strategy 1	30	40	35	105
Strategy 2	35	60	90	185
Total	65	100	125	290

TABLE 20
Expected Frequencies of the Number of Sales for Two Different Strategies During a 3-Month Period

Month	1	2	3	Total
Strategy 1	23.5	36.2	45.3	105
Strategy 2	41.5	63.8	79.7	185
Total	65	100	125	290

frequency distribution shows that the two measures are *not* independent, they must be distributed in a systematic fashion with respect to the criterion.

Let us add another variable to the previous drug sales example. The variable is two different advertising strategies from two different pharmacies. Table 19 presents the frequencies of the number of sales during the same 3-month period according to two different advertising strategies.

The expected frequencies are calculated from the formula

$$E = \frac{R * C}{N}$$

where E = the expected frequency in a particular cell, R = the sum of the frequencies in a particular row, C = the sum of the frequencies in a particular column, and N = the sum of the frequencies for all cells. The formula shows that the expected cell frequencies are computed by multiplying the row frequency by the column frequency for any given cell, then dividing this product by the overall total frequency. For example, the expected frequency of strategy 1 in the first month is $105 \times 65 / 290 = 23.5$. The expected frequency of strategy 2 in the second month is computed as $185 \times 100 / 290 = 63.8$. The expected frequencies for each cell can be computed by the same method and are presented in Table 20.

The appropriate hypotheses are as follows:

- H_0: The advertising strategy and the number of drug sales are independent.
- H_1: The advertising strategy and the number of drug sales are not independent.

The chi-square allows us to test the hypothesis, and the calculations are demonstrated as follows:

$$\chi^2 = (30 - 23.5)^2/23.5 + (40 - 36.2)^2/36.2 + (35 - 45.3)^2/45.3$$
$$+ (35 - 41.5)^2/41.5 + (60 - 63.8)^2/63.8 + (90 - 79.7)^2/79.7$$
$$= 7.0$$

The degrees of freedom for the two-way classifications are computed as

$$df = (\text{row} - 1)(\text{column} - 1) = (2 - 1)(3 - 1) = 2$$

The critical value of chi-square with df = 2 is 5.99 ($\alpha = 0.05$). Since the computed value of chi-square (7.0) is greater than 5.99, we reject the null hypothesis and conclude that the advertising strategies and the number of drug sales are related ($p < 0.05$). While the chi-square test of independence does not specifically identify the difference between frequencies of the two advertising methods, an examination of the observed frequencies from the contingency table suggests that strategy 2 shows an increasing number of drug sales over the 3-month period while strategy 1 does not show a substantial change.

A word of caution is appropriate regarding small expected frequencies. When a small sample size and a large number of categories are used in a study, some cells may yield small or even zero expected frequencies. The distribution of chi-square is not strictly valid with small expected frequencies. There is a debate, however, regarding what constitutes small expected frequencies. Cochran[4] suggests that no expected frequency should be below 1 if no more than 20% of the cells have expected frequencies of less than 5. A more conservative view is that no expected frequency should be below 6.

The solution to the small expected frequencies problem is to combine adjacent rows and/or columns to meet the minimum expected frequency, as long as such combining does not violate the conceptual structure of the classification scheme. In addition to the small expected frequencies problem, researchers should be aware of the small sample size problem. If the sample size is less than 10, the chi-square distribution is not an appropriate sampling distribution for the chi-square statistic.

The chi-square only indicates whether the variables are independent or related — whether at least one of the observed frequencies is significantly different from its expected frequency. It does not describe the strength of the relationship. A relationship may be statistically significant without being *practically* significant. There are several statistics, however, that do measure the strength of the relationship. Interested readers are referred to the SPSS manual.[14]

In spite all the above problems, chi-square analysis, which can apply to ordinal as well as nominal data, is easily and widely applicable to test hypotheses of independence in a contingency table. If more than two variables are involved it is difficult to interpret the results. In this case, a more advanced statistical technique called log-linear models for the analysis of categorical data is useful. Like multiple regression, log-linear models use all classification variables as the independent variables and the number of cases in a cell of the classification as the dependent variables. The interaction among independent variables can also be examined. The discussion of log-linear models is beyond the scope of this book. Interested readers may read the books by Haberman[8] and Bishop et al.[2].

3. Concluding Remarks

Descriptive statistics serve to describe and summarize data. Measures of central tendency and variability provide indices to represent the average value of a whole set of scores and the spread of the scores within a distribution.

Inferential statistics enable us to estimate the population parameters from the phenomena observed in samples. The most commonly used inferential statistics are the t-test, analysis of variance, regression analysis, and the chi-square test. The t-test is used to determine whether the difference between two means is significant. Analysis of variance is used to compare the means of two or more samples. Regression analysis describes the mathematical relationship between a dependent variable and one or more independent variables. The chi-square test is used to find the differences between proportions of subjects or events in different categories, by comparing observed frequencies and expected frequencies.

REFERENCES — SECTION II

1. **Anderson, R. and Bancroft, T.,** *Statistical Theory in Research,* McGraw-Hill, New York, 1952.
2. **Bishop, Y. M. M., Feinberg, S. E., and Holland, P. W.,** *Discrete Multivariate Analysis: Theory and Practice,* MIT Press, Cambridge, MA, 1975.
3. **Blalock, H. M., Jr.,** *Social Statistics,* 2nd ed., McGraw-Hill, New York, 1972.

4. **Cochran, W. G.,** Some methods for strengthening the common χ tests, *Biometrics,* 10, 315, 1954.
5. **Cronbach, L. J.,** Coefficient alpha and the internal structure of tests, *Psychometrica,* 16, 297, 1951.
6. **Glass, G. V. and Stanley, J. C.,** *Statistical Methods in Education and Psychology,* Prentice-Hall, Englewood Cliffs, NJ, 1970.
7. **Greenhouse, S. W. and Geisser, S.,** On methods in the analysis of profile data, *Psychometrica,* 24, 95, 1959.
8. **Haberman, S. J.,** *Analysis of Qualitative Data,* Vol. 1, Academic Press, London, 1978.
9. **Hartley, H. O.,** The maximum F-ratio as a short-cut test for heterogeneity of variance, *Biometrika,* 37, 308, 1950.
10. **Kachigan, S. K.,** *Multivariate Statistical Analysis,* Radius Press, New York, 1982.
11. **Kirk, R.,** *Experimental Design: Procedures for the Behavioral Sciences,* Brooks, Belmont, CA, 1968.
12. **Lancaster, H. O.,** *The Chi-Square Distribition,* John Wiley & Sons, New York, 1969.
13. **Neter, J. and Wasserman, W.,** *Applied Linear Statistical Models,* Richard D. Irwin, Homewood, IL, 1974.
14. **Nie, N. H., Hull, C. H., Jenkins, J. G., Steinbrenner, K., and Bent, D. H.,** *Statistical Packages for the Social Sciences,* 2nd ed., McGraw-Hill, New York, 1975.
15. **Nunnally, J. C.,** *Psychometric Theory,* McGraw-Hill, New York, 1978.
16. **Poor, D. D.,** Analysis of variance for repeated measures designs: two approaches, *Psychol. Bull.,* 80, 204, 1973.
17. **Richardson, G. F. and Kuder, M.,** The theory of the estimation of test reliability, *Psychometrika,* 2, 135, 1937.
18. **Roscoe, J. T.,** *Fundamental Research Statistics for the Behavioral Sciences,* Holt, Rinehart & Winston, New York, 1969.
19. **SPSSx (1986).** *SPSSx User's Guide,* 2nd ed., SPSS Inc., Chicago, IL, 1986.
20. **Winer, B. J.,** *Statistical Principles in Experimental Design,* 2nd ed., McGraw-Hill, New York, 1971.

III. GUIDE TO SELECTING APPROPRIATE STATISTICAL METHODS IN ANALYZING RESEARCH STUDIES

The choice of the statistical analysis to be used in analyzing a health research study should be a part of the research planning. The decision on the appropriate use and type of statistical analysis should be made prior to data collection, not after the fact. In deciding on an appropriate method the following should be considered, but the choice should not be limited to these questions/statements.

1. The nature of the research study, i.e., description, exploratory, explanatory, hypothesis testing etc.
2. What are you trying to find?
3. How will samples be obtained?
4. Type of sample and sampling procedure to be used
5. Type of data to be generated for the research
6. Are you interested in one, two, or more variables?
7. Are you dealing with one, two or more than two sample problems?

Answering these questions during the research design will help a researcher to identify the appropriate statistical technique for data analysis.

The major types of statistical analysis as well as their appropriate use will be discussed briefly to aid the health researcher in choosing appropriate statistical techniques for their research data.

Hypothesis testing provides a basis for decision making on an objective rather than a subjective basis by weighing the relative probabilities of different hypotheses by simply looking at the data. Different people will form different opinions from a clinical uniform criterion for making consistent decisions. Hypothesis testing may be in one, two, or more sample inferences. Where the clinical research is interested in one-variable and one-sample problems, see Figure 16 for statistical analyses that can be used depending on the situation. Figures 17 and 18 deal with one-variable two-sample and more than two-sample problems, respectively. Figures 19 and 20 are concerned with research studies interested in relationships between two or more variables, respectively. Oftentimes in hypothesis testing our population variance is unknown. When a researcher is faced with such a situation, the *t*-test is used in the hypothesis testing. The *t*-test can be one tailed or two tailed (sometimes called one or two sided, respectively) depending on what the researcher is testing. The primary advantage of the *t*-statistic is that one need not know the population parameters to conduct a *t*-test. Many researchers, therefore, use the *t*-test whenever the population parameters are not known, since the *t*-distribution can be used for all sample sizes and takes the sample size into account.

Chi-square uses a measure of association based on true statistical independence. It is used to compare a set of observed frequencies obtained for the research study and expected frequencies based on the hypothesis or theoretical speculation which is independent of the observed data. Chi-square, therefore, represents a useful method of comparing experimentally obtained results with those to be expected theoretically on the hypothesis.

In a 2×2 contingency table or for chi-square with one degree of freedom it is common practice to use the Yates correction to produce an adjusted value of χ^2. However, not all authors are in favor of using the Yates correction.[10]

McNemar's test is used where the observed data is not independent, e.g., obtaining two or more sets of data from the same patient. For example, this test could be used for measuring blood levels of a drug on the same patient with different methods in an effort to evaluate the different methods.

There is no unanimous agreement on the minimum size of the expected frequency for the use of chi-square. It has been suggested that it should not be less than 5.[11] Another suggestion is that no expected frequency should be less than 1 and not over 20% of the expected frequency should be less than 5.[12] The problem of small expected frequency can be solved by combining categories as long as it is done appropriately. It will be inappropriate to combine categories in an attempt to increase or decrease the value of χ^2. In the case of small expected frequencies, Fisher's Exact Test should be used instead of chi-square.

Chi-square tests the existence of association between the variables in question; it does not estimate the degree of the association.

The analysis of variance is used when testing for differences between three or more means simultaneously. It tests if there are any differences among the means of different populations from which various samples have been drawn. There are certain assumptions underlying the use of analysis of variance:

1. Each sample is a random sample from the corresponding population.
2. The samples are independent of each other.
3. The sample comes from a normally distributed population.
4. The populations have similar variance (homoscedasticity).
5. The variable being measured is distributed normally in the population.

In analysis of variance, the F-test is used to see if the differences between the sample means could be due to variation within each group instead of differences between the groups being studied. Analysis of variance can be one-way (i.e., studying one dependent variable under different conditions), or more than one-way analysis of variance may be used where a researcher is studying the effectiveness of different formulations of antipyretic drugs in reducing fever. In this instance, the dependent variables are (1) the type of antipyretic drug

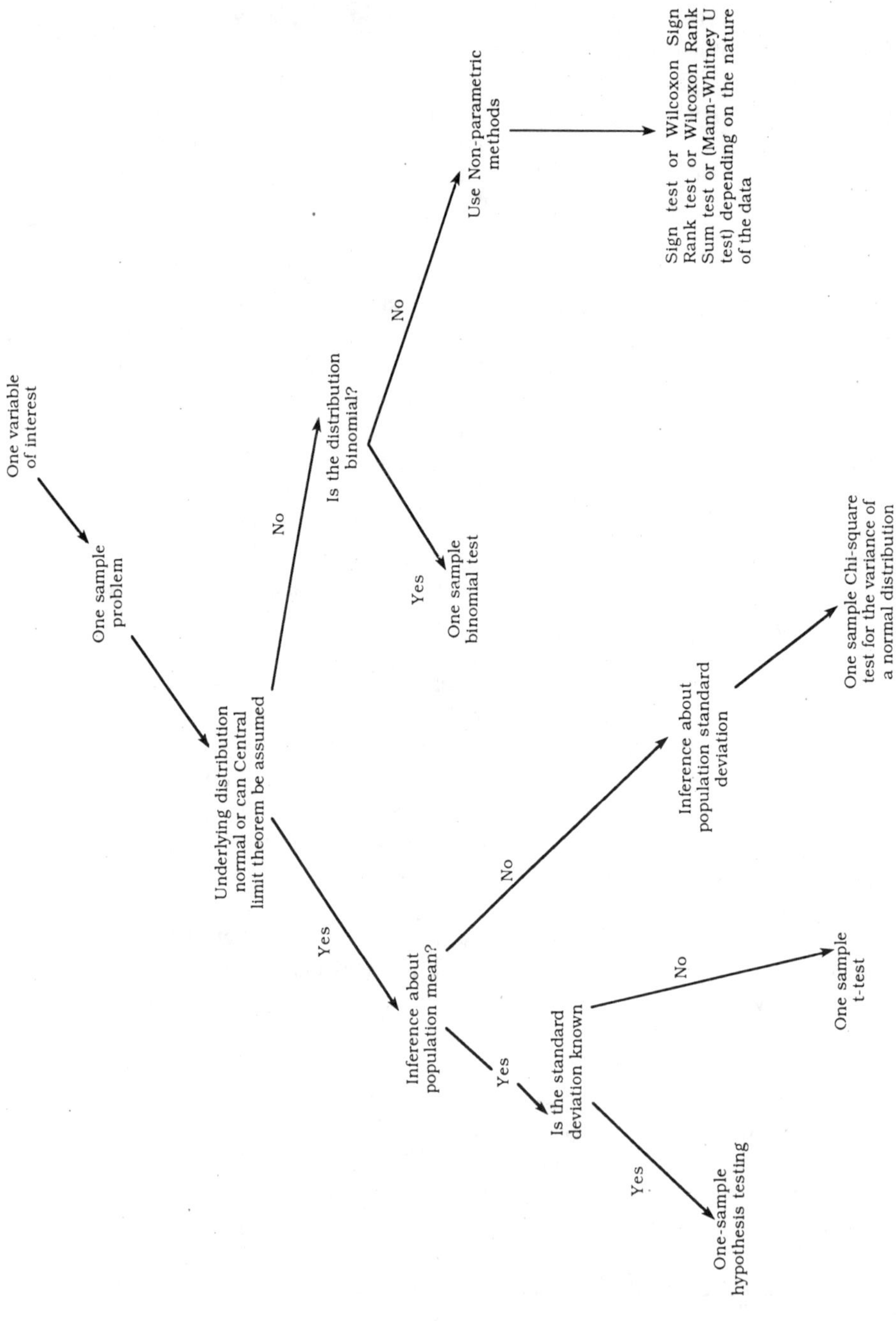

FIGURE 16. One-variable, one-group sample problem.

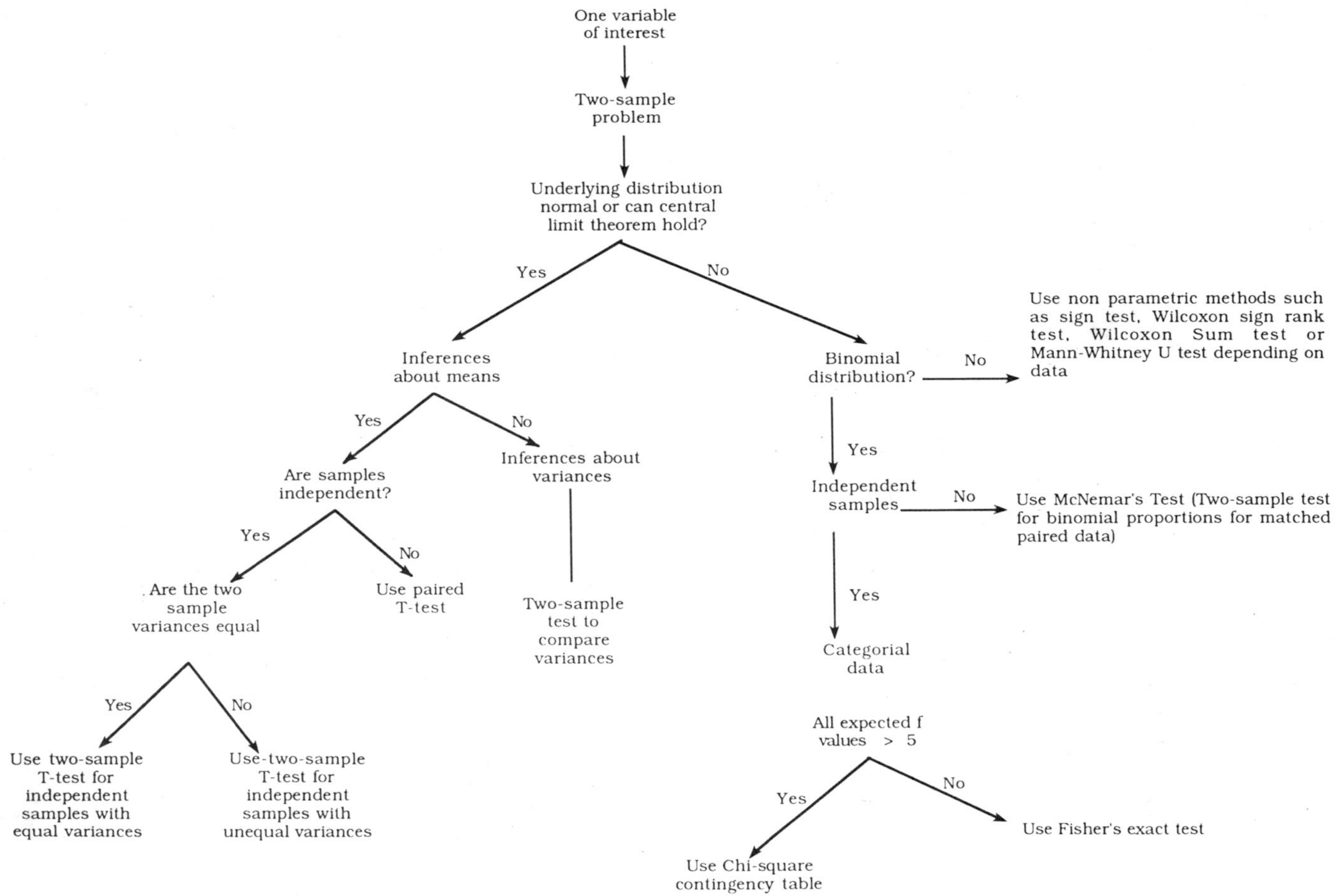

FIGURE 17. One-variable, two-group sample problem.

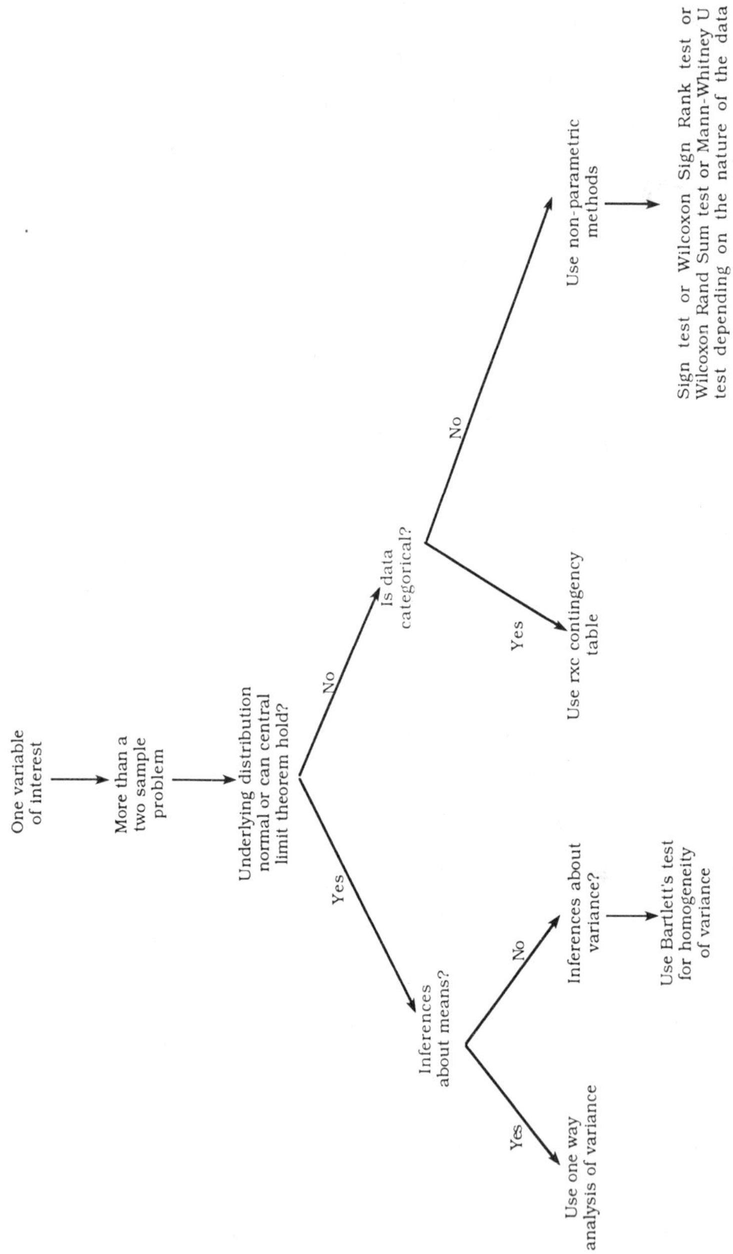

FIGURE 18. One-variable, multiple-group sample problem.

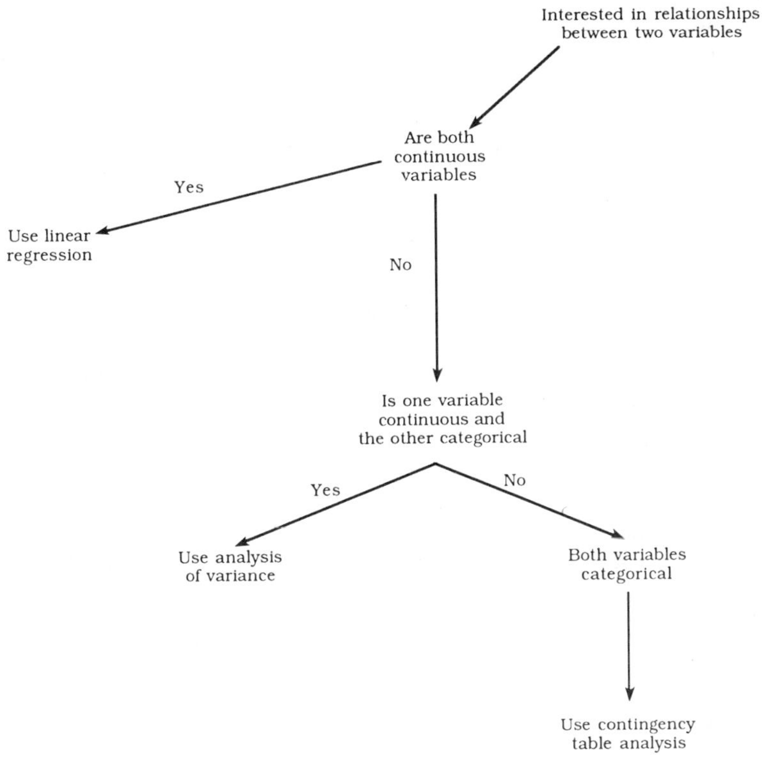

FIGURE 19. Two-variable problem.

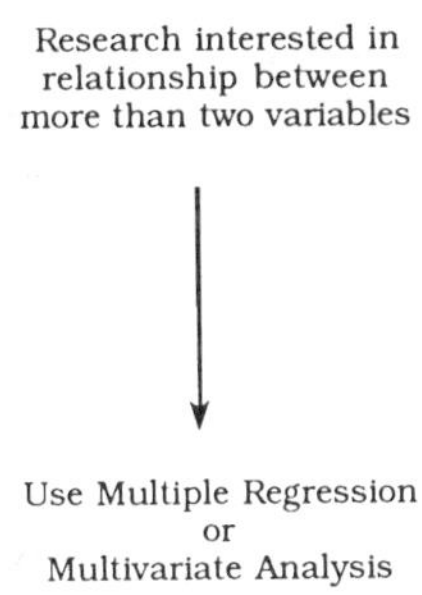

FIGURE 20. Multiple-variable problem.

and (2) the different formulations. These variables may have several levels; e.g., six different formulations of four different drugs may be analyzed simultaneously.

Regression and correlation are concerned with association among measurements of different variables. Regression analysis is a means of studying the variation of the dependent variable at different levels of the independent variable. Thus, it is possible to estimate or predict the value of a dependent variable on the basis of a known measurement of an independent variable. Simple linear regression deals with problems involving linear relationships between one dependent and one independent variable. Multilinear regression is used in cases involving one dependent and more than one independent variable.

Correlation is used to measure the degree or strength of relationships between variables. The linear correlation coefficient is a measure of the linear relationship between two random variables. The most commonly used measure of linear correlation between two variables is the Pearson product-moment correlation coefficient (r) or, simply, the simple correlation coefficient. The sign of r indicates the direction of the relationship; it has nothing to do with its strength. It is the value of r that indicates the degree or strength of the relationship. Pearson's is best suited for linear relationships for variables that are both continuous and quantitative in nature, e.g., interval and ratio data. There are other ways of measuring association, such as partial correlation, multiple correlation, rank order correlation (ρ) (when data are in the form of rank order), correlation ratio (η) (for curvilinear relationships), phi coefficient (ϕ), contingency coefficient, biserial correlation, point biserial correlation, and tetrachoric correlation. For more information on nonlinear regression and correction, please refer to the works of Hays[13] and Blalock.[14]

Some of the nonparametric methods of data analysis are indicated in Figures 16 to 18. These methods will be discussed briefly. The sign test is used where the clinical research data is qualitative assessment (e.g., more redness or more rash) for two dependent samples, such as paired data. However, the pairs should be independent of one another. The Wilcoxon sign rank test is used where data is obtained in a quantitative assessment, e.g., degree of burn. The data obtained is from two dependent samples (i.e., paired data), but the pairs should also be independent. The Wilcoxon rank sum test or Mann-Whitney U test is a nonparametric analogue to the *t*-test for two independent samples. This test is nonparametric because it is based on the ranks of the individual observation or on the relative magnitude of the data collected.

In conclusion, some of the basic statistical methods of data analysis have been presented. Figures 16 to 20 are presented to serve as a guide for the researcher in choosing appropriate statistical analyses based on their type of research studies. Statistics are excellent tools in research data analysis; however, where inappropriately used they may make the results of a well-conducted research study uninterpretable or meaningless.

REFERENCES — SECTION III

1. **Kruskal, William H.,** Statistics: the field, in *International Encyclopedia of the Social Sciences,* Vol. 15, Sils, D. L., Ed., Macmillan, New York, 1968, 206.
2. **Remington, R. D. and Schork, M. A.,** *Statistics with Applications to the Biological and Health Sciences,* Prentice-Hall, Englewood Cliffs, NJ, 1970.
3. **Glass, G. and Stanley, J. C.,** *Statistical Methods in Education and Psychology,* Prentice-Hall, Englewood Cliffs, NJ, 1970.
4. **Ferguson, G. A.,** *Statistical Analysis in Psychology and Education,* McGraw-Hill, New York, 1976.
5. **Rosner, B.,** *Fundamentals of Biostatistics,* Duxbury Press, Boston, 1982.
6. **Garrett, H. E. and Woodworth, R. S.,** *Statistics in Psychology and Education,* David McKay, New York, 1971.
7. **Bahn, A. K.,** *Basic Medical Statistics,* Grune & Stratton, New York, 1980.
8. **Hawkins, C. A. and Weber, J. E.,** *Statistical Analysis Application to Business and Economics,* Harper & Row, New York, 1980.
9. **Huff, D.,** *How to Lie With Statistics,* W. W. Norton, New York, 1954.
10. **Grizzle, J. E.,** Continuity correction in the χ^2 test for 2 × 2 tables, *Am. Stat.,* 21, 28, 1967.
11. **Garrett, H. E. and Woodworth, R. S.,** *Statistics in Psychology and Education,* David McKay, New York, 1971.
12. **Remington, R. D. and Schork, M. A.,** *Statistics with Application to the Biological and Health Sciences,* Prentice-Hall, Englewood Cliffs, NJ, 1970.
13. **Hays, W. R.,** *Statistics for the Social Sciences,* Holt, Rinehart & Winston, New York, 1973, 675.
14. **Blalock, H. M.,** *Social Statistics,* McGraw-Hill, New York, 1972, 408.
15. **Simon, R.,** Confidence intervals for reporting results of clinical trials, *Ann. Int. Med.,* 105(3), 429, 1986.
16. **Wolf, F. M. et al.,** Interpreting behavioral, biomedical and psychological relationships in chronic disease from 2 × 2 tables using correlation, *J. Chronic Dis.,* 39(8), 605, 1986.
17. **Clayden, A. D.,** Assessing your research readiness, *Practitioner,* 230(1417), 643, 1986.
18. **Hauck, W. W. et al.,** A proposal for interpreting and reporting negative studies, *Stat. Med.,* 5(3), 203, 1986.

APPENDIX

TABLE 1
Areas under the Normal Curve

z	Second decimal place in z									
	0.00	**0.01**	**0.02**	**0.03**	**0.04**	**0.05**	**0.06**	**0.07**	**0.08**	**0.09**
0.0	0.0000	0.0040	0.0080	0.0120	0.0160	0.0199	0.0239	0.0279	0.0319	0.0359
0.1	0.0398	0.0438	0.0478	0.0517	0.0557	0.0596	0.0636	0.0675	0.0714	0.0753
0.2	0.0793	0.0832	0.0871	0.0910	0.0948	0.0987	0.1026	0.1064	0.1103	0.1141
0.3	0.1179	0.1217	0.1255	0.1293	0.1331	0.1368	0.1406	0.1443	0.1480	0.1517
0.4	0.1554	0.1591	0.1628	0.1664	0.1700	0.1736	0.1772	0.1808	0.1844	0.1879
0.5	0.1915	0.1950	0.1985	0.2019	0.2054	0.2088	0.2123	0.2157	0.2190	0.2224
0.6	0.2257	0.2291	0.2324	0.2357	0.2389	0.2422	0.2454	0.2486	0.2517	0.2549
0.7	0.2580	0.2611	0.2642	0.2673	0.2704	0.2734	0.2764	0.2794	0.2823	0.2852
0.8	0.2881	0.2910	0.2939	0.2967	0.2995	0.3023	0.3051	0.3078	0.3106	0.3133
0.9	0.3159	0.3186	0.3212	0.3238	0.3264	0.3289	0.3315	0.3340	0.3365	0.3389
1.0	0.3413	0.3438	0.3461	0.3485	0.3508	0.3531	0.3554	0.3577	0.3599	0.3621
1.1	0.3643	0.3665	0.3686	0.3708	0.3729	0.3749	0.3770	0.3790	0.3810	0.3830
1.2	0.3849	0.3869	0.3888	0.3907	0.3925	0.3944	0.3962	0.3980	0.3997	0.4015
1.3	0.4032	0.4049	0.4066	0.4082	0.4099	0.4115	0.4131	0.4147	0.4162	0.4177
1.4	0.4192	0.4207	0.4222	0.4236	0.4251	0.4265	0.4279	0.4292	0.4306	0.4319
1.5	0.4332	0.4345	0.4357	0.4370	0.4382	0.4394	0.4406	0.4418	0.4429	0.4441
1.6	0.4452	0.4463	0.4474	0.4484	0.4495	0.4505	0.4515	0.4525	0.4535	0.4545
1.7	0.4554	0.4564	0.4573	0.4582	0.4591	0.4599	0.4608	0.4616	0.4625	0.4633
1.8	0.4641	0.4649	0.4656	0.4664	0.4671	0.4678	0.4686	0.4693	0.4699	0.4706
1.9	0.4713	0.4719	0.4726	0.4732	0.4738	0.4744	0.4750	0.4756	0.4761	0.4767
2.0	0.4772	0.4778	0.4783	0.4788	0.4793	0.4798	0.4803	0.4808	0.4812	0.4817
2.1	0.4821	0.4826	0.4830	0.4834	0.4838	0.4842	0.4846	0.4850	0.4854	0.4857
2.2	0.4861	0.4864	0.4868	0.4871	0.4875	0.4878	0.4881	0.4884	0.4887	0.4890
2.3	0.4893	0.4896	0.4898	0.4901	0.4904	0.4906	0.4909	0.4911	0.4913	0.4916
2.4	0.4918	0.4920	0.4922	0.4925	0.4927	0.4929	0.4931	0.4932	0.4934	0.4936
2.5	0.4938	0.4940	0.4941	0.4943	0.4945	0.4946	0.4948	0.4949	0.4951	0.4952
2.6	0.4953	0.4955	0.4956	0.4957	0.4959	0.4960	0.4961	0.4962	0.4963	0.4964
2.7	0.4965	0.4966	0.4967	0.4968	0.4969	0.4970	0.4971	0.4972	0.4973	0.4974
2.8	0.4974	0.4975	0.4976	0.4977	0.4977	0.4978	0.4979	0.4979	0.4980	0.4981
2.9	0.4981	0.4982	0.4982	0.4983	0.4984	0.4984	0.4985	0.4985	0.4986	0.4986
3.0	0.4987	0.4987	0.4987	0.4988	0.4988	0.4989	0.4989	0.4989	0.4990	0.4990
3.1	0.4990	0.4991	0.4991	0.4991	0.4992	0.4992	0.4992	0.4992	0.4993	0.4993
3.2	0.4993	0.4993	0.4994	0.4994	0.4994	0.4994	0.4994	0.4995	0.4995	0.4995
3.3	0.4995	0.4995	0.4995	0.4996	0.4996	0.4996	0.4996	0.4996	0.4996	0.4997
3.4	0.4997	0.4997	0.4997	0.4997	0.4997	0.4997	0.4997	0.4997	0.4997	0.4998
3.5	0.4998									
4.0	0.49997									
4.5	0.499997									
5.0	0.4999997									

Note: Proportion of total area under the normal curve between mean ordinate and ordinate at given z distance from the mean.

TABLE 2
Critical Values of the Pearson Correlation Coefficient

	Level of significance for one-tailed test			
	0.05	**0.025**	**0.01**	**0.005**
	Level of significance for the two-tailed test			
df	**0.10**	**0.05**	**0.02**	**0.01**
1	0.988	0.997	0.9995	0.9999
2	0.900	0.950	0.980	0.990
3	0.805	0.878	0.934	0.959
4	0.729	0.811	0.882	0.917
5	0.669	0.754	0.833	0.874
6	0.622	0.707	0.789	0.834
7	0.582	0.666	0.750	0.798
8	0.549	0.632	0.716	0.765
9	0.521	0.602	0.685	0.735
10	0.497	0.576	0.658	0.708
11	0.576	0.553	0.634	0.684
12	0.458	0.532	0.612	0.661
13	0.441	0.514	0.592	0.641
14	0.426	0.497	0.574	0.623
15	0.412	0.482	0.558	0.606
16	0.400	0.468	0.542	0.590
17	0.389	0.456	0.528	0.575
18	0.378	0.444	0.516	0.561
19	0.369	0.433	0.503	0.549
20	0.360	0.423	0.492	0.537
21	0.352	0.413	0.482	0.526
22	0.344	0.404	0.472	0.515
23	0.337	0.396	0.462	0.505
24	0.330	0.388	0.453	0.496
25	0.323	0.381	0.445	0.487
26	0.317	0.374	0.437	0.479
27	0.311	0.367	0.430	0.471
28	0.306	0.361	0.423	0.463
29	0.301	0.355	0.416	0.486
30	0.296	0.349	0.409	0.449
35	0.275	0.325	0.381	0.418
40	0.257	0.304	0.358	0.393
45	0.243	0.288	0.338	0.372
50	0.231	0.273	0.322	0.354
60	0.211	0.250	0.295	0.325
70	0.195	0.232	0.274	0.303
80	0.183	0.217	0.256	0.283
90	0.173	0.205	0.242	0.267
100	0.164	0.195	0.230	0.254

Adapted from Fisher, R. A. and Yates, F., *Statistical Tables for Biological, Agricultural, and Medical Research,* Oliver & Boyd, Edinburgh, 1974. With permission.

TABLE 3
Critical Values of the Spearman Correlation Coefficient

N	Significance level (one-tailed test) 0.05	0.01
4	1.000	
5	0.900	1.000
6	0.829	0.943
7	0.714	0.893
8	0.643	0.833
9	0.600	0.783
10	0.564	0.746
12	0.506	0.712
14	0.456	0.645
16	0.425	0.601
18	0.399	0.564
20	0.377	0.534
22	0.359	0.508
24	0.343	0.485
26	0.329	0.465
28	0.317	0.448
30	0.306	0.432

Adapted from Olds, E. G., *Ann. Math. Stat.*, 9, 133, 1938; *Ann. Math. Stat.*, 20, 117, 1949. With permission.

TABLE 4
Distribution of *t* for Given Probability Levels

	Level of significance for one-tailed test					
	0.10	0.05	0.025	0.01	0.005	0.0005
	Level of significance for two-tailed test					
df	0.20	0.10	0.05	0.02	0.01	0.001
1	3.078	6.314	12.706	31.821	63.657	636.619
2	1.886	2.920	4.303	6.965	9.925	31.598
3	1.638	2.353	3.182	4.541	5.841	12.941
4	1.533	2.132	2.776	3.747	4.604	8.610
5	1.476	2.015	2.571	3.365	4.032	6.859
6	1.440	1.943	2.447	3.143	3.707	5.959
7	1.415	1.895	2.365	2.998	3.499	5.405
8	1.397	1.860	2.306	2.896	3.355	5.041
9	1.383	1.833	2.262	2.821	3.250	4.781
10	1.372	1.812	2.228	2.764	3.169	4.587
11	1.363	1.796	2.201	2.718	3.106	4.437
12	1.356	1.782	2.179	2.681	3.055	4.318
13	1.350	1.771	2.160	2.650	3.012	4.221
14	1.345	1.761	2.145	2.624	2.977	4.140
15	1.341	1.753	2.131	2.602	2.947	4.073
16	1.337	1.746	2.120	2.583	2.921	4.015
17	1.333	1.740	2.110	2.567	2.898	3.965
18	1.330	1.734	2.101	2.552	2.878	3.992
19	1.328	1.729	2.093	1.539	2.861	3.883
20	1.325	1.725	2.086	2.528	2.845	3.850
21	1.323	1.721	2.080	2.518	2.831	3.819
22	1.321	1.717	2.074	2.508	2.819	3.792
23	1.319	1.714	2.069	2.500	2.807	3.767
24	1.318	1.711	2.064	2.492	2.797	3.745
25	1.316	1.708	2.060	2.485	2.787	3.725
26	1.315	1.706	2.056	2.479	2.779	3.707
27	1.314	1.703	2.052	2.473	2.771	3.690
28	1.313	1.701	2.048	2.467	2.763	3.674
29	1.311	1.699	2.045	2.462	2.756	3.659
30	1.310	1.697	2.042	2.457	2.750	3.646
40	1.303	1.684	2.021	2.423	2.704	3.551
60	1.296	1.671	2.000	2.390	2.660	3.460
120	1.289	1.658	1.980	2.358	2.617	3.373
∞	1.282	1.645	1.960	2.326	2.576	3.291

Adapted from Fisher, R. A. and Yates, F., *Statistical Tables for Biological, Agricultural, and Medical Research,* Oliver and Boyd, Edinburgh, 1974. With permission.

TABLE 5
Upper 5% and 1% Points of F_{max} = (Largest s^2)/(Smallest s^2) for k Samples from Normal Populations, Each Providing v Degrees of Freedom for s^2

k / v	2	3	4	5	6	7	8	9	10	11	12
						5% points					
2	39.0	87.5	142	202	266	333	403	475	550	626	704
3	15.4	27.8	39.2	50.7	62.0	72.9	83.5	93.9	104	114	124
4	9.60	15.5	20.6	25.2	29.5	33.6	37.5	41.1	44.6	48.0	51.4
5	7.15	10.8	13.7	16.3	18.7	20.8	22.9	24.7	26.5	28.2	29.9
6	5.82	8.38	10.4	12.1	13.7	15.0	16.3	17.5	18.6	19.7	20.7
7	4.99	6.94	8.44	9.70	10.8	11.8	12.7	13.5	14.3	15.1	15.8
8	4.43	6.00	7.18	8.12	9.03	9.78	10.5	11.1	11.7	12.2	12.7
9	4.03	5.34	6.31	7.11	7.80	8.41	8.95	9.45	9.91	10.3	10.7
10	3.72	4.85	5.67	6.34	6.92	7.42	7.87	8.28	8.66	9.01	9.34
12	3.28	4.16	4.79	5.30	5.72	6.09	6.42	6.72	7.00	7.25	7.48
15	2.86	3.54	4.01	4.37	4.68	4.95	5.19	5.40	5.59	5.77	5.93
20	2.46	2.95	3.29	3.54	3.76	3.94	4.10	4.24	4.37	4.49	4.59
30	2.07	2.40	2.61	2.78	2.91	3.02	3.12	3.21	3.29	3.36	3.39
60	1.67	1.85	1.96	2.04	2.11	2.17	2.22	2.26	2.30	2.33	2.36
∞	1.00	1.00	1.00	1.00	1.00	1.00	1.00	1.00	1.00	1.00	1.00
						1% points					
2	199	448	729	1036	1362	1705	2063	2432	2813	3204	3605
3	47.5	85	120	151	184	21(6)	24(9)	28(1)	31(0)	33(7)	36(1)
4	23.2	37	49	59	69	79	89	97	106	113	120
5	14.9	22	28	33	38	42	46	50	54	57	60
6	11.1	15.5	19.1	22	25	27	30	32	34	36	37
7	8.89	12.1	14.5	16.5	18.4	20	22	23	24	26	27
8	7.50	9.9	11.7	13.2	14.5	15.8	16.9	17.9	18.9	19.8	21
9	6.54	8.5	9.9	11.1	12.1	13.1	13.9	14.7	15.3	16.0	16.6
10	5.85	7.4	8.6	9.6	10.4	11.1	11.8	12.4	12.9	13.4	13.9
12	4.91	6.1	6.9	7.6	8.2	8.7	9.1	9.5	9.9	10.2	10.6
15	4.07	4.9	5.5	6.0	6.4	6.7	7.1	7.3	7.5	7.8	8.0
20	3.32	3.8	4.3	4.6	4.9	5.1	5.3	5.5	5.6	5.8	5.9
30	2.63	3.0	3.3	3.4	3.6	3.7	3.8	3.9	4.0	4.1	4.2
60	1.96	2.2	2.3	2.4	2.4	2.5	2.5	2.6	2.6	2.7	2.7
∞	1.00	1.0	1.0	1.0	1.0	1.0	1.0	1.0	1.0	1.0	1.0

Note: Values in the column $k = 2$ and in the rows $v = 2$ and $v = \infty$ are exact. Elsewhere the third digit may be in error by a few units for the 5% points and several units for the 1% points. The third-digit figures in parentheses for $v = 3$ are the most uncertain.

From Hartley, H. O., *Biometrika,* 39, 424, 1952. With permission.

TABLE 6
The *F*-Distribution (0.05 Level)

	df$_1$																		
df$_2$	**1**	**2**	**3**	**4**	**5**	**6**	**7**	**8**	**9**	**10**	**12**	**15**	**20**	**24**	**30**	**40**	**60**	**120**	**∞**
1	161.4	199.5	215.7	224.6	230.2	234.0	236.8	238.9	240.5	241.9	243.9	245.9	248.0	249.1	250.1	251.1	252.2	253.3	254.3
2	18.51	19.00	19.16	19.25	19.30	19.33	19.35	19.37	19.38	19.40	19.41	19.43	19.45	19.45	19.46	19.47	19.48	19.49	19.50
3	10.13	9.55	9.28	9.12	9.01	8.94	8.89	8.85	8.81	8.79	8.74	8.70	8.66	8.64	8.62	8.59	8.57	8.55	8.53
4	7.71	6.94	6.59	6.39	6.26	6.16	6.09	6.04	6.00	5.96	5.91	5.86	5.80	5.77	5.75	5.72	5.69	5.66	5.63
5	6.61	5.79	5.41	5.19	5.05	4.95	4.88	4.82	4.77	4.74	4.68	4.62	4.56	4.53	4.50	4.46	4.43	4.40	4.36
6	5.99	5.14	4.76	4.53	4.39	4.28	4.21	4.15	4.10	4.06	4.00	3.94	3.87	3.84	3.81	3.77	3.74	3.70	3.67
7	5.59	4.74	4.35	4.12	3.97	3.87	3.79	3.73	3.68	3.64	3.57	3.51	3.44	3.41	3.38	3.34	3.30	3.27	3.23
8	5.32	4.46	4.07	3.84	3.69	3.58	3.50	3.44	3.39	3.35	3.28	3.22	3.15	3.12	3.08	3.04	3.01	2.97	2.93
9	5.12	4.26	3.86	3.63	3.48	3.37	3.29	3.23	3.18	3.14	3.07	3.01	2.94	2.90	2.86	2.83	2.79	2.75	2.71
10	4.96	4.10	3.71	3.48	3.33	3.22	3.14	3.07	3.02	2.98	2.91	2.85	2.77	2.74	2.70	2.66	2.62	2.58	2.54
11	4.84	3.98	3.59	3.36	3.20	3.09	3.01	2.95	2.90	2.85	2.79	2.72	2.65	2.61	2.57	2.53	2.49	2.45	2.40
12	4.75	3.89	3.49	3.26	3.11	3.00	2.91	2.85	2.80	2.75	2.69	2.62	2.54	2.51	2.47	2.43	2.38	2.34	2.30
13	4.67	3.81	3.41	3.18	3.03	2.92	2.83	2.77	2.71	2.67	2.60	2.53	2.46	2.42	2.38	2.34	2.30	2.25	2.21
14	4.60	3.74	3.34	3.11	2.96	2.85	2.76	2.70	2.65	2.60	2.53	2.46	2.39	2.35	2.31	2.27	2.22	2.18	2.13
15	4.54	3.68	3.29	3.06	2.90	2.79	2.71	2.64	2.59	2.54	2.48	2.40	2.33	2.29	2.25	2.20	2.16	2.11	2.07
16	4.49	3.63	3.24	3.01	2.85	2.74	2.66	2.59	2.54	2.49	2.42	2.35	2.28	2.24	2.19	2.15	2.11	2.06	2.01
17	4.45	3.59	3.20	2.96	2.81	2.70	2.61	2.55	2.49	2.45	2.38	2.31	2.23	2.19	2.15	2.10	2.06	2.01	1.96
18	4.41	3.55	3.16	2.93	2.77	2.66	2.58	2.51	2.46	2.41	2.34	2.27	2.19	2.15	2.11	2.06	2.02	1.97	1.92
19	4.38	3.52	3.13	2.90	2.74	2.63	2.54	2.48	2.42	2.38	2.31	2.23	2.16	2.11	2.07	2.03	1.98	1.93	1.88
20	4.35	3.49	3.10	2.87	2.71	2.60	2.51	2.45	2.39	2.35	2.28	2.20	2.12	2.08	2.04	1.99	1.95	1.90	1.84
21	4.32	3.47	3.07	2.84	2.68	2.57	2.49	2.42	2.37	2.32	2.25	2.18	2.10	2.05	2.01	1.96	1.92	1.87	1.81
22	4.30	3.44	3.05	2.82	2.66	2.55	2.46	2.40	2.34	2.30	2.23	2.15	2.07	2.03	1.98	1.94	1.89	1.84	1.78
23	4.28	3.42	3.03	2.80	2.64	2.53	2.44	2.37	2.32	2.27	2.20	2.13	2.05	2.01	1.96	1.91	1.86	1.81	1.76
24	4.26	3.40	3.01	2.78	2.62	2.51	2.42	2.36	2.30	2.25	2.18	2.11	2.03	1.98	1.94	1.89	1.84	1.79	1.73
25	4.24	3.39	2.99	2.76	2.60	2.49	2.40	2.34	2.28	2.24	2.16	2.09	2.01	1.96	1.92	1.87	1.82	1.77	1.71
26	4.23	3.37	2.98	2.74	2.59	2.47	2.39	2.32	2.27	2.22	2.15	2.07	1.99	1.95	1.90	1.85	1.80	1.75	1.69
27	4.21	3.35	2.96	2.73	2.57	2.46	2.37	2.31	2.25	2.20	2.13	2.06	1.97	1.93	1.88	1.84	1.79	1.73	1.67
28	4.20	3.34	2.95	2.71	2.56	2.45	2.36	2.29	2.24	2.19	2.12	2.04	1.96	1.91	1.87	1.82	1.77	1.71	1.65
29	4.18	3.33	2.93	2.70	2.55	2.43	2.35	2.28	2.22	2.18	2.10	2.03	1.94	1.90	1.85	1.81	1.75	1.70	1.64
30	4.17	3.32	2.92	2.69	2.53	2.42	2.33	2.27	2.21	2.16	2.09	2.01	1.93	1.89	1.84	1.79	1.74	1.68	1.62
40	4.08	3.23	2.84	2.61	2.45	2.34	2.25	2.18	2.12	2.08	2.00	1.92	1.84	1.79	1.74	1.69	1.64	1.58	1.51

60	4.00	3.15	2.76	2.53	2.37	2.25	2.17	2.10	2.04	1.99	1.92	1.84	1.75	1.70	1.65	1.59	1.53	1.47	1.39
120	3.92	3.07	2.68	2.45	2.29	2.17	2.09	2.02	1.96	1.91	1.83	1.75	1.66	1.61	1.55	1.50	1.43	1.35	1.25
∞	3.84	3.00	2.60	2.37	2.21	2.10	2.01	1.94	1.88	1.83	1.75	1.67	1.57	1.52	1.46	1.39	1.32	1.22	1.00

0.01 Level

1	4052	4999.5	5403	5625	5764	5859	5928	5982	6022	6056	6106	6157	6209	6235	6261	6287	6313	6339	6366
2	98.5	99.00	99.17	99.25	99.30	99.33	99.36	99.37	99.39	99.40	99.42	99.43	99.45	99.46	99.47	99.47	99.48	99.49	99.50
3	34.12	30.82	29.46	28.71	28.24	27.91	27.67	27.49	27.35	27.23	27.05	26.87	26.69	26.60	26.50	26.41	26.32	26.22	26.13
4	21.20	18.00	16.69	15.98	15.52	15.21	14.98	14.80	14.66	14.55	14.37	14.20	14.02	13.93	13.84	13.75	13.65	13.56	13.46
5	16.26	13.27	12.06	11.39	10.97	10.67	10.46	10.29	10.16	10.05	9.89	9.72	9.55	9.47	9.38	9.29	9.20	9.11	9.02
6	13.75	10.92	9.78	9.15	8.75	8.47	8.26	8.10	7.98	7.87	7.72	7.56	7.40	7.31	7.23	7.14	7.06	6.97	6.88
7	12.25	9.55	8.45	7.85	7.46	7.19	6.99	6.81	6.72	6.62	6.47	6.31	6.16	6.07	5.99	5.91	5.82	5.74	5.65
8	11.26	8.65	7.59	7.01	6.63	6.37	6.18	6.03	5.91	5.81	5.67	5.52	5.36	5.28	5.20	5.12	5.03	4.95	4.86
9	10.56	8.02	6.99	6.42	6.06	5.80	5.61	5.47	5.35	5.26	5.11	4.96	4.81	4.73	4.65	4.57	4.48	4.40	4.31
10	10.04	7.56	6.55	5.99	5.64	5.39	5.20	5.06	4.94	4.85	4.71	4.56	4.41	4.33	4.25	4.17	4.08	4.00	3.91
11	9.65	7.21	6.22	5.67	5.32	5.07	4.89	4.74	4.63	4.54	4.40	4.25	4.10	4.02	3.94	3.86	3.78	3.69	3.60
12	9.33	6.93	5.95	5.41	5.06	4.82	4.64	4.50	4.39	4.30	4.16	4.01	3.86	3.78	3.70	3.62	3.54	3.45	3.36
13	9.07	6.70	5.74	5.21	4.86	4.62	4.44	4.30	4.19	4.10	3.96	3.82	3.66	3.59	3.51	3.43	3.34	3.25	3.17
14	8.86	6.51	5.56	5.04	4.69	4.46	4.28	4.14	4.03	3.94	3.80	3.66	3.51	3.43	3.35	3.27	3.18	3.09	3.00
15	8.68	6.36	5.42	4.89	4.56	4.32	4.14	4.00	3.89	3.80	3.67	3.52	3.37	3.29	3.21	3.13	3.05	2.96	2.87
16	8.53	6.23	5.29	4.77	4.44	4.20	4.03	3.89	3.78	3.69	3.55	3.41	3.26	3.18	3.10	3.02	2.93	2.84	2.75
17	8.40	6.11	5.18	4.67	4.34	4.10	3.93	3.79	3.68	3.59	3.46	3.31	3.16	3.08	3.00	2.92	2.83	2.75	2.65
18	8.29	6.01	5.09	4.58	4.25	4.01	3.84	3.71	3.60	3.51	3.37	3.23	3.08	3.00	2.92	2.84	2.75	2.66	2.57
19	8.18	5.93	5.01	4.50	4.17	3.94	3.77	3.63	3.52	3.43	3.30	3.15	3.00	2.92	2.84	2.76	2.67	2.58	2.49
20	8.10	5.85	4.94	4.43	4.10	3.87	3.70	3.56	3.46	3.37	3.23	3.09	2.94	2.86	2.78	2.69	2.61	2.52	2.42
21	8.02	5.78	4.87	4.37	4.04	3.81	3.64	3.51	3.40	3.31	3.17	3.03	2.88	2.80	2.72	2.64	2.55	2.46	2.36
22	7.95	5.72	4.82	4.31	3.99	3.76	3.59	3.45	3.35	3.26	2.12	2.98	2.83	2.75	2.67	2.58	2.50	2.40	2.31
23	7.88	5.66	4.76	4.26	3.94	3.71	3.54	3.41	3.30	3.21	3.07	2.93	2.78	2.70	2.62	2.54	2.45	2.35	2.26
24	7.82	5.61	4.72	4.22	3.90	3.67	3.50	3.36	3.26	3.17	3.03	2.89	2.74	2.66	2.58	2.49	2.40	2.31	2.21
25	7.77	5.57	4.68	4.18	3.85	3.63	3.46	3.32	3.22	3.13	2.99	2.85	2.70	2.62	2.54	2.45	2.36	2.27	2.17
26	7.72	5.53	4.64	4.14	3.82	3.59	3.42	3.29	3.18	3.09	2.96	2.81	2.66	2.58	2.50	2.42	2.33	2.23	2.13
27	7.68	5.49	4.60	4.11	3.78	3.56	3.39	3.26	3.15	3.06	2.93	2.78	2.63	2.55	2.47	2.38	2.29	2.20	2.10
28	7.64	5.45	4.57	4.07	3.75	3.53	3.36	3.23	3.12	3.03	2.90	2.75	2.60	2.52	2.44	2.35	2.26	2.17	2.06
29	7.60	5.42	4.54	4.04	3.73	3.50	3.33	3.20	3.09	3.00	2.87	2.73	2.57	2.49	2.41	2.33	2.23	2.14	2.03
30	7.56	4.39	4.51	4.02	3.70	3.47	3.30	3.17	3.07	2.98	2.84	2.70	2.55	2.47	2.39	2.30	2.21	2.11	2.01
40	7.31	5.18	4.31	3.83	3.51	3.29	3.12	2.99	2.89	2.80	2.66	2.52	2.37	2.29	2.20	2.11	2.02	1.92	1.80
60	7.08	4.98	4.13	3.65	3.34	3.12	2.95	2.82	2.72	2.63	2.50	2.35	2.20	2.12	2.03	1.94	1.84	1.73	1.60

TABLE 6 (continued)
The F-Distribution (0.05 Level)

df_2	df_1: 1	2	3	4	5	6	7	8	9	10	12	15	20	24	30	40	60	120	∞
120	6.85	4.79	3.95	3.48	3.17	2.96	2.79	2.66	2.56	2.47	2.34	2.19	2.03	1.95	1.86	1.76	1.66	1.53	1.38
∞	6.63	4.61	3.78	3.32	3.02	2.80	2.64	2.51	2.41	2.32	2.18	2.04	1.88	1.79	1.70	1.59	1.47	1.32	1.00

$1 - \alpha = 0.95$

	r: 2	3	4	5	6	7	8	9	10	11	12	13	14	15	16	17	18	19	20
1	18.0	27.0	32.8	37.1	40.4	43.1	45.4	47.4	49.1	50.6	52.0	53.2	54.3	55.4	56.3	57.2	58.0	58.8	59.6
2	6.08	8.33	9.80	10.9	11.7	12.4	13.0	13.5	14.0	14.4	14.7	15.1	15.4	15.7	15.9	16.1	16.4	16.6	16.8
3	4.50	5.91	6.82	7.50	8.04	8.48	8.85	9.18	9.46	9.72	9.95	10.2	10.3	10.5	10.7	10.8	11.0	11.1	11.2
4	3.93	5.04	5.76	6.29	6.71	7.05	7.35	7.60	7.83	8.03	8.21	8.37	8.52	8.66	8.79	8.91	9.03	9.13	9.23
5	3.64	4.60	5.22	5.67	6.03	6.33	6.58	6.80	6.99	7.17	7.32	7.47	7.60	7.72	7.83	7.93	8.03	8.12	8.21
6	3.46	4.34	4.90	5.30	5.63	5.90	6.12	6.32	6.49	6.65	6.79	6.92	7.03	7.14	7.24	7.34	7.43	7.51	7.59
7	3.34	4.16	4.68	5.06	5.36	5.61	5.82	6.00	6.16	6.30	6.43	6.55	6.66	6.76	6.85	6.94	7.02	7.10	7.17
8	3.26	4.04	4.53	4.89	5.17	5.40	5.60	5.77	5.92	6.05	6.18	6.29	6.39	6.48	6.57	6.65	6.73	6.80	6.87
9	3.20	3.95	4.41	4.76	5.02	5.24	5.43	5.59	5.74	5.87	5.98	6.09	6.19	6.28	6.36	6.44	6.51	6.58	6.64
10	3.15	3.88	4.33	4.65	4.91	5.12	5.30	5.46	5.60	5.72	5.83	5.93	6.03	6.11	6.19	6.27	6.34	6.40	6.47
11	3.11	2.82	4.26	4.57	4.82	5.03	5.20	5.35	5.49	5.61	5.71	5.81	5.90	5.98	6.06	6.13	6.20	6.27	6.33
12	3.08	3.77	4.20	4.51	4.75	4.95	5.12	5.27	5.39	5.51	5.61	5.71	5.80	5.88	5.95	6.02	6.09	6.15	5.21
13	3.06	3.73	4.15	4.45	4.69	4.88	5.05	5.19	5.32	5.43	5.53	5.63	5.71	5.79	5.86	5.93	5.99	6.05	6.11
14	3.03	3.70	4.11	4.41	4.64	4.83	4.99	5.13	5.25	5.36	5.46	5.55	5.64	5.71	5.79	5.85	5.91	5.97	6.03
15	3.01	3.67	4.08	4.37	4.59	4.78	4.94	5.08	5.20	5.31	5.40	5.49	5.57	5.65	5.72	5.78	5.85	5.90	5.96
16	3.00	3.65	4.05	4.33	4.56	4.74	4.90	5.03	5.15	5.26	5.35	5.44	5.52	5.59	5.66	5.73	5.79	5.84	5.90
17	2.98	3.63	4.02	4.30	4.52	4.70	4.86	4.99	5.11	5.21	5.31	5.39	5.47	5.54	5.61	5.67	5.73	5.79	5.84
18	2.97	3.61	4.00	4.28	4.49	4.67	4.82	4.96	5.07	5.17	5.27	5.35	5.43	5.50	5.57	5.63	5.69	5.74	5.79
19	2.96	3.59	3.98	4.25	4.47	4.65	4.79	4.92	5.04	5.14	5.23	5.31	5.39	5.46	5.53	5.59	5.65	5.70	5.75
20	2.95	3.58	3.96	4.23	4.45	4.62	4.77	4.90	5.01	5.11	5.20	5.28	5.36	5.43	5.49	5.55	5.61	5.66	5.71
24	2.92	3.53	3.90	4.17	4.37	4.54	4.68	4.81	4.92	5.01	5.10	5.18	5.25	5.32	5.38	5.44	5.49	5.55	5.59

30	2.89	3.49	3.85	4.10	4.30	4.46	4.60	4.72	4.82	4.92	5.00	5.08	5.15	5.21	5.27	5.33	5.38	5.43	5.47
40	2.86	3.44	3.79	4.04	4.23	4.39	4.52	4.63	4.73	4.82	4.90	4.98	5.04	5.11	5.16	5.22	5.27	5.31	5.36
60	2.83	3.40	3.74	3.98	4.16	4.31	4.44	4.55	4.65	4.73	4.81	4.88	4.94	5.00	5.06	5.11	5.15	5.20	5.24
120	2.80	3.36	3.68	3.92	4.10	4.24	4.36	4.47	4.56	4.64	4.71	4.78	4.84	4.90	4.95	5.00	5.04	5.09	5.13
∞	2.77	3.31	3.63	3.86	4.03	4.17	4.29	4.39	4.47	4.55	4.62	4.68	4.74	4.80	4.85	4.89	4.93	4.97	5.01

Adapted from Pearson, E. S. and Huntley, H. O., Eds., *Biometrika Tables for Statisticians,* Vol. 1, 1st ed., Cambridge University Press, London, 1962, 179. With permission.

TABLE 7
Percentiles of the Studentized Range Distribution

$1 - \alpha = 0.99$

ν	r: 2	3	4	5	6	7	8	9	10	11	12	13	14	15	16	17	18	19	20
1	90.0	135	164	186	202	216	227	237	246	253	260	266	272	277	282	286	290	294	298
2	14.0	19.0	22.3	24.7	26.6	28.2	29.5	30.7	31.7	32.6	33.4	34.1	34.8	35.4	36.0	36.5	37.0	37.5	37.9
3	8.26	10.6	12.2	13.3	14.2	15.0	15.6	16.2	16.7	17.1	17.5	17.9	18.2	18.5	18.8	19.1	19.3	19.5	19.8
4	6.51	8.12	9.17	9.96	10.6	11.1	11.5	11.9	12.3	12.6	12.8	13.1	13.3	13.5	13.7	13.9	14.1	14.2	14.4
5	5.70	6.97	7.80	8.42	8.91	9.32	9.67	9.97	10.2	10.5	10.7	10.9	11.1	11.2	11.4	11.6	11.7	11.8	11.9
6	5.24	6.33	7.03	7.56	7.97	8.32	8.61	8.87	9.10	9.30	9.49	9.65	9.81	9.95	10.1	10.2	10.3	10.4	10.5
7	4.95	5.92	6.54	7.01	7.37	7.68	7.94	8.17	8.37	8.55	8.71	8.86	9.00	9.12	9.24	9.35	9.46	9.55	9.65
8	4.74	5.63	6.20	6.63	6.96	7.24	7.47	7.68	7.87	8.03	8.18	8.31	8.44	8.55	8.66	8.76	8.85	8.94	9.03
9	4.60	5.43	5.96	6.35	6.66	6.91	7.13	7.32	7.49	7.65	7.78	7.91	8.03	8.13	8.23	8.32	8.41	8.49	8.57
10	4.48	5.27	5.77	6.14	6.43	6.67	6.87	7.05	7.21	7.36	7.48	7.60	7.71	7.81	7.91	7.99	8.07	8.15	8.22
11	4.39	5.14	5.62	5.97	6.25	6.48	6.67	6.84	6.99	7.13	7.25	7.36	7.46	7.56	7.65	7.73	7.81	7.88	7.95
12	4.32	5.04	5.50	5.84	6.10	6.32	6.51	6.67	6.81	6.94	7.06	7.17	7.26	7.36	7.44	7.52	7.59	7.66	7.73
13	4.26	4.96	5.40	5.73	5.98	6.19	6.37	6.53	6.67	6.79	6.90	7.01	7.10	7.19	7.27	7.34	7.42	7.48	7.55
14	4.21	4.89	5.32	5.63	5.88	6.08	6.26	6.41	6.54	6.66	6.77	6.87	6.96	7.05	7.12	7.20	7.27	7.33	7.39
15	4.17	4.83	5.25	5.56	5.80	5.99	6.16	6.31	6.44	6.55	6.66	6.76	6.84	6.93	7.00	7.07	7.14	7.20	7.26
16	4.13	4.78	5.19	5.49	5.72	5.92	6.08	6.22	6.35	6.46	6.56	6.66	6.74	6.82	6.90	6.97	7.03	7.09	7.15
17	4.10	4.74	5.14	5.43	5.66	5.85	6.01	6.15	6.27	6.38	6.48	6.57	6.66	6.73	6.80	6.87	6.94	7.00	7.05
18	4.07	4.70	5.09	5.38	5.60	5.79	5.94	6.08	6.20	6.31	6.41	6.50	6.58	6.65	6.72	6.79	6.85	6.91	6.96
19	4.05	4.67	5.05	5.33	5.55	5.73	5.89	6.02	6.14	6.25	6.34	6.43	6.51	6.58	6.65	6.72	6.78	6.84	6.89
20	4.02	4.64	5.02	5.29	5.51	5.69	5.84	5.97	6.09	6.19	6.29	6.37	6.45	6.52	6.59	6.65	6.71	6.76	6.82
24	3.96	4.54	4.91	5.17	5.37	5.54	5.69	5.81	5.92	6.02	6.11	6.19	6.26	6.33	6.39	6.45	6.51	6.56	6.61
30	3.89	4.45	4.80	5.05	5.24	5.40	5.54	5.65	5.76	5.85	5.93	6.01	6.08	6.14	6.20	6.26	6.31	6.36	6.41
40	3.82	4.37	4.70	4.93	5.11	5.27	5.39	5.50	5.60	5.69	5.77	5.84	5.90	5.96	6.02	6.07	6.12	6.17	6.21
60	3.76	4.28	4.60	4.82	4.99	5.13	5.25	5.36	5.45	5.53	5.60	5.67	5.73	5.79	5.84	5.89	5.93	5.98	6.02
120	3.70	4.20	4.50	4.71	4.87	5.01	5.12	5.21	5.30	5.38	5.44	5.51	5.56	5.61	5.66	5.71	5.75	5.79	5.83
∞	3.64	4.12	4.40	4.60	4.76	4.88	4.99	5.08	5.16	5.23	5.29	5.35	5.40	5.45	5.49	5.54	5.57	5.61	5.65

From Scheffé, H., *The Analysis of Variance*, John Wiley & Sons, New York, 1959, 434. With permission.

TABLE 8
Percentiles of the Chi-Square Distribution

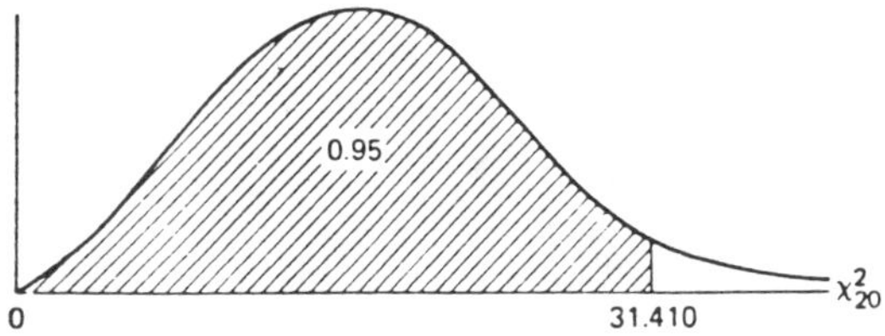

df	$\chi^2_{0.005}$	$\chi^2_{0.025}$	$\chi^2_{0.05}$	$\chi^2_{0.90}$	$\chi^2_{0.95}$	$\chi^2_{0.975}$	$\chi^2_{0.99}$	$\chi^2_{0.995}$
1	0.0000393	0.000982	0.00393	2.706	3.841	5.024	6.635	7.879
2	0.0100	0.0506	0.103	4.605	5.991	7.378	9.210	10.597
3	0.0717	0.216	0.352	6.251	7.815	9.348	11.345	12.838
4	0.207	0.484	0.711	7.779	9.488	11.143	13.277	14.860
5	0.412	0.831	1.145	9.236	11.070	12.832	15.086	16.750
6	0.676	1.237	1.635	10.645	12.592	14.449	16.812	18.548
7	0.989	1.690	2.167	12.017	14.067	16.013	18.475	20.278
8	1.344	2.180	2.733	13.362	15.507	17.535	20.090	21.955
9	1.735	2.700	3.325	14.684	16.919	19.023	21.666	23.589
10	2.156	3.247	3.940	15.987	18.307	20.483	23.209	25.188
11	2.603	3.816	4.575	17.275	19.675	21.920	24.725	26.757
12	3.074	4.404	5.226	18.549	21.026	23.336	26.217	28.300
13	3.565	5.009	5.892	19.812	22.362	24.736	27.688	29.819
14	4.075	5.629	6.571	21.064	23.685	26.119	29.141	31.319
15	4.601	6.262	7.261	22.307	24.996	27.488	30.578	32.801
16	5.142	6.908	7.962	23.542	26.296	28.845	32.000	34.267
17	5.697	7.564	8.672	24.769	27.587	30.191	33.409	35.718
18	6.265	8.231	9.390	25.989	28.869	31.526	34.805	37.156
19	6.844	8.907	10.117	27.204	30.144	32.852	36.191	38.582
20	7.434	9.591	10.851	28.412	31.410	34.170	37.566	39.997
21	8.034	10.283	11.591	29.615	32.671	35.479	38.932	41.401
22	8.643	10.982	12.338	30.813	33.924	36.781	40.289	42.796
23	9.260	11.688	13.091	32.007	35.172	38.076	41.638	44.181
24	9.886	12.401	13.848	33.196	36.415	39.364	42.980	45.558
25	10.520	13.120	14.611	34.382	37.652	40.646	44.314	46.928
26	11.160	13.844	15.379	35.563	38.885	41.923	45.642	48.290
27	11.808	14.573	16.151	36.741	40.113	43.194	46.963	49.645
28	12.461	15.308	16.928	37.916	41.337	44.461	48.278	50.993
29	13.121	16.047	17.708	39.087	42.557	45.722	49.588	52.336
30	13.787	16.791	18.493	40.256	43.773	46.979	50.892	53.672
35	17.192	20.569	22.465	46.059	49.802	53.203	57.342	60.275
40	20.707	24.433	26.509	51.805	55.758	59.342	63.691	66.766
45	24.311	28.366	30.612	57.505	61.656	65.410	69.957	73.166
50	27.991	32.357	34.764	63.167	67.505	71.420	76.154	79.490
60	35.535	40.482	43.188	74.397	79.082	83.298	88.379	91.952
70	43.275	48.758	51.739	85.527	90.531	95.023	100.425	104.215
80	51.172	57.153	60.391	96.578	101.879	106.629	112.329	116.321
90	59.196	65.647	69.126	107.565	113.145	118.136	124.116	128.299
100	67.328	74.222	77.929	118.498	124.342	129.561	135.807	140.169

Chapter 5

INTERPRETING AND REPORTING RESEARCH FINDINGS

J. Fincham and E. Nnadi Okolo

TABLE OF CONTENTS

* J. Fincham.
** E. Nnadi Okolo.

I. INTERPRETING RESULTS

The interpretation of the results of studies and research projects is the culminating process in any program of research. The investigator at this stage has the possibility of reaching, through publication, many individuals who would not otherwise know of a particular research study or project. The opportunity to learn from one's own research as well as describe a project to countless others requires a suitable interpretation of the data derived from the research project.

A. DATA INTERPRETATION

Authors must have a firm understanding of their research results in order to adequately express to others their findings. Chances are that if the researcher does not fully understand his/her research findings, others will not be able to follow the author's interpretation of the research findings. It will help to organize the data in sequential fashion for an orderly presentation.

Regardless of the degree of sophistication of the data analysis techniques utilized, the researcher should begin the interpretive process by examining the demographic makeup of the study sample. In addition, the descriptive statistics derived from the study should be examined. Crucial questions must be asked at this stage of the research process; they may include the following:

1. Does the sample size or selection criteria for the sample introduce a bias that must be addressed before and during data interpretation?
2. Does the statistical test utilized seem appropriate when considering assumptions that might have been violated in the sample selection process?
3. Was the statistical test utilized robust (as is analysis of variance) to violations of assumptions?
4. When considering the interpretation of the findings are there other factors that must be considered?

Others will be better able to follow an interpretation of results fluently if it is presented in an orderly fashion.

When interpreting results the author should fully describe the data set to the reader. The reader should be allowed to examine both the numerical values for a variable as well as the verbal translation of the value. For example, if using a five-point Likert scale with anchors of

- Strongly agree (numerical value = 1)
- Agree (numerical value = 2)
- Uncertain (numerical value = 3)
- Disagree (numerical value = 4)
- Strongly disagree (numerical value = 5)

the numerical values should not be listed solely by themselves. A verbal anchor should accompany the description of the variable. Consider the following hypothetical example.

In an attitudinal survey a group of researchers were interested in examining the attitudes of health professional students concerning their upcoming graduation. One of the statements in the survey was as follows: "I am confident I will be able to make a difference in the delivery of services to my patients." The sample was asked to rate this and other questions on a five-point Likert scale as described above. The analysis of the responses to this particular item indicated a mean score of 1.2 with a standard deviation of 0.62 as the overall sample response to the statement. If the value of 1.2 with a standard deviation of 0.62 was reported in the write-up, the readers may or may not have had an understanding of the meaning of the value. However, consider if in addition to the listing of the value the author made the following statement: "This numerical value can be translated verbally from the Likert format to indicate that, on average, the sample members tended to strongly agree that they would be able to make a difference in the delivery of services to patients." The reader would have more of a feel for how the author interpreted the item and response. Certainly readers could disagree with the interpretation, but the author has at least described an interpretation of the responses, and this allows the reader to make an assessment of the interpretation.

Thus, the interpretive process should begin with a description of the data on an elementary level. After this component has been completed the more elaborate statistical measures or analyses should be presented.

B. INFERENCE

When considering the process of inferring from the sample to the population, conservative responses are called for. Before inferring from a sample to the population a careful assessment of the sample representativeness should be undertaken. Does the sample adequately portray the population from which it was derived? If the sample does appear to be representative, then statements can be made generalizing from the results in the study sample to the population in general. However, the word "caution" cannot be overutilized in this process.

Say, for the sake of an example, a new treatment for nicotine addiction was tested in a sample of smoking patients. The sample utilized was a sample of patients attending a mental health clinic of a large health maintenance organization (HMO). The sample was made up of 50 males, 25 of whom were assigned to a treatment group and received the test drug. The other 25 patients were assigned to the no treatment control group. At the conclusion of the study the treatment group was found to have a 30% increase in smoking cessation in comparison with the control group. What should be inferred from this sample to a population? The results were significant; however, is the sample makeup suspect? Should the results be inferred to all smoking patients, both male and female? Does the fact that the patients were drawn from a sample of mental health clinic patients contained in an HMO influence how the results should be related to the population at large? If the study was carried out with a sample of smokers which reflected the population as a whole, then inferential statements relating to the population of smokers could be made more appropriately.

When implying an inferential association between a sample and a population seems appropriate for assessment of results, the level of significance should be listed numerically. The reader should have the opportunity to examine the results and assess the degree of association between the findings regardless of whether significance or nonsignificance was found. To simply state "statistical significance was found" leaves the reader without the degree of knowledge that should be possessed pertaining to the findings.

C. PREDICTION

The use of the term "prediction" should be considered carefully. It is difficult to predict anything in any arena, including research. However, the use of theory in research does afford the author the ability to deal with prediction. Kerlinger[1] has noted the following: "If A, then B, A and B referring to classes of objects or events." This does imply prediction, and one of the advantages of theory-based research is the opportunity to utilize theory to achieve a certain degree of prediction. It should be kept in mind that results in the one setting derived from theory-based research are not necessarily indicative of results in other settings or in all cases. If a multivariate technique such as multiple regression or discriminant analysis, which more appropriately deal with prediction of variance than other statistical tests, is utilized, a certain degree of prediction can be alluded to. However, the statistical assumptions must be addressed before comments relative to predictability are espoused.

D. SIGNIFICANCE AND NONSIGNIFICANCE

One of the most misused terms in the language of research is the word "significance". An equally misused term is "nonsignificance". Significance has traditionally been defined as a probability ≤ 0.05, and highly significant results are listed with a probability of ≤ 0.01. A probability value of 0.05 indicates that the results derived could be the result of chance in 5 out of 100 opportunities. The value of 0.01 is indicative of a chance occurrence in 1 out of 100 opportunities.

Researchers will in some cases speak of significant or highly significant results without listing the actual probability value derived. It should be routine practice in the interpretation of statistical results to list the actual probability value determined and let the reader make an informed decision about the results after examination of the actual value derived from the research analysis. If a p value of 0.051 was found in a statistical analysis and the author spoke of the results being nonsignificant, one should question the appropriateness of the discussion. By the same token, if a researcher spoke of the "significance" of a p value of 0.049, the discussion also should be scrutinized closely.

In classical hypothesis testing, "*a priori*" criteria of significance (acceptance or rejection of a null hypothesis) are put in place. If an acceptance value of 0.05 is set in advance, theoretically a value of 0.051 would lead to the acceptance of the null hypothesis, and a value of 0.049 would lead to the nonrejection of the null hypothesis. If classical hypothesis testing is utilized and an *a priori* acceptance value of $\alpha = 0.05$ or 0.01 is chosen, acceptance or rejection of a null hypothesis is appropriate. However, the author should also address the issue of practical vs. statistical significance of the findings.

E. REPETITIVE DATA ANALYSIS PROCEDURES

Another issue relevant for researchers revolves around the number of statistical runs to which a data set is exposed. If a data set is analyzed 20 times and one statistically significant result is found, this would correspond to a finding of significance 5% of the time. The point of this discussion is that caution is called for in interpretating results found to be significant after many computer runs of the same data set. This caution should not preclude a researcher from fully interpreting the findings of a research study. Close scrutiny from many angles is appropriate and worthwhile. The point simply is that the more computer runs a data set is put through, the more likely is the possibility of finding statistical significance.

F. CAUSALITY

Prudence is called for in considering the construct of causality. Issues pertaining to the use of the term "causality" must be addressed conservatively. An association that is found to be significant does not imply that the association automatically leads to causality. Statistically significant relationships do not imply that one variable is causing the variation in another variable or series of variables.

Causation certainly cannot be "proved" through the finding of statistically significant relationships. Consider, for an example, that two variables are found to be correlated at a level of $p = 0.80$ in a Pearson's product moment correlation analysis (Pearson's r). Let us assume for the sake of example that the two variables are education and income. The correlation is positive in nature, such that income and education show parallel increases in the sample studied. When an author would speak of the relationship between these two variables, mention could be made of the variables predicting 64% of the variance in the other variable ($[r = 0.8]^2 =$ percent of variance explained). However, addressing the notion of causality in relation to these two variables would not be possible. Does a high level of education "cause" a high income? Conversely, does a high income "cause" a high level of education? Another reason to *not* consider using the term causality when exploring the relationship between these two variables is that a third variable or series of variables may be affecting the parallel increase in income and education. The identity of the variable may be unknown, or perhaps it may be considered to be implicitly known, but immeasurable.

An unknown variable is just that — a variable which the researcher neglects or ignores. The end result is the same — the impact of the variable is not considered in the analysis and interpretation of the data. Perhaps an implicit variable is "opportunity" or "position". A study sample may be made up of individuals in the same company with apparent similar job advancement opportunities or having the benefit of an opportunity to have attended a certain college or university. However, even if the variable is indeed implicit, the individuals' responses to the opportunity may vary and influence the variable or variables of interest. Achievement, motivation, or equal opportunity might erroneously be considered equal for all individuals in the sample. However, individual variable responses, internal drive, or biases toward certain individuals by management (racism or sexism) may make a "controlled" implicit variable actually perform as a manipulated variable which then impacts upon the two variables in question.

Causality is something all researchers would perhaps seek to find, and in some disciplines or types of studies it may be more reachable than in others. For instance, medical researchers examining the cause of disease or the effectiveness of a drug treatment may have more of a grasp for determining effects of variables than do behavioral researchers. However, every worthy research endeavor contains ethereal impacting variables that are not considered and/or measured. Something all scientists share when considering causality is the fact that, regardless of discipline or field of endeavor, causality is a construct that is beyond the absolute reach of researchers. If all variables were always known by researchers, would research be as enjoyable or fruitful as it is? Do not the uncertain aspects of the "unknowns" of research actually drive the investigator to pursue the unknown with a fervor?

G. REPORTING OF RESULTS INDEPENDENT OF SIGNIFICANCE FOUND

Regarding interpretation, results of well-conducted studies should be reported regardless of findings of significance or nonsignificance. Significant findings can include significant or nonsignificant results. There appears to be a bias toward the reporting of research results where significant findings are found. Results which indicate a degree of statistical significance should be published; science is advanced through the reporting of such results. However, nonsignificant results found in studies that are well planned and well conducted also need to be published. Begg and Berlin[2] have suggested that this publication bias has introduced many false-positive results in the literature. They further suggest that the false positives can have a serious impact on the comparison of research results. The technique of meta-analysis, which compares the results of many studies and has as an output one standardized statistical value, would seem to be influenced quantitatively in a negative fashion by publication bias. All evaluative aspects of research, whether quantitative or qualitative, would seem to be influenced negatively by publication bias.

The goal of research should be to further knowledge of a particular facet of the unknown. Research is, after all, the systematic study and analysis of the unknown. Because of this, nonsignificant findings need to be reported as well. If only significant findings of research studies are published, a selection bias is in place which impedes the advancement of a growing body of knowledge in a certain discipline or field. Nonsignificant statistics obtained from a research analysis may add to a body of the knowledge of great practical significance. Certainly the same can be stated concerning statistically significant results. However, statistically significant results can produce practically insignificant knowledge. If findings are practically significant and add to an accumulating knowledge base pertinent to a field of study, they should be published regardless of statistical significance achieved.

When considering statistical significance vs. practical significance it may be useful to ask the following question: "Does the reporting of values explain my results?" If, for example, variables are correlated at a level of $r = 0.10$, with a p value <0.001, in a sample size of 1000, a total of 1% of the variance is being explained. The practical significance of this value may supercede the statistical significance achieved. Small values are important when adding something new to the research literature or the knowledge base of science. For example, consider a predictive study of longevity in which researchers studied the impact of various variables on life span. These variables included cholesterol, weight, sodium intake, or cardiovascular disease. If a total of 1% added variance in life-span determination could be determined by a relationship, even in a large sample size with a small-appearing correlation, the practical significance of the determination would outweigh or be equally important to the statistical significance level achieved.

Regarding interpretation, an r level >0.70 is considered meaningful, an r level <0.50 is, in fact, explaining $<25\%$ of the variation, and an r value <0.20 is explaining $<4\%$ of the variation in the variables in question. The interpretation of research reports is just as important as the actual collection of data. Researchers would do well to focus much attention on the appropriate interpretive response to collected data.

II. REPORTING RESULTS

A. TYPES OF WRITE-UPS

There are five types of formalized write-ups from research activities: case reports, commentaries, position papers, literature reviews, and research reports. Certainly any of these specific types of reports are not necessarily mutually exclusive. However, for the sake of simplicity, it would be useful to consider each as a separate entity. A given requirement for each of the following types of written reports of research findings is an adequate and thorough documentation of previous or related work in the literature.

1. Case Report

Case reports are descriptions of single isolated cases of either managerial, clinical, or financial content. The case report usually contains detailed information regarding the case in question. This full description allows the reader to follow the description and progress of a specific detailed example. Because of the single-case specificity of case reports they have a lack of generalization to other patients, situations, circumstances, etc. Unless fully documented, case reports can be more anecdotal than informational. Nevertheless, they do serve as a publishing mechanism to share information that is useful to other researchers, clinicians, managers, etc. Venulet[3] has suggested criteria for case report publication. Even though the initial application of Venulet's criteria was for the reporting of adverse drug reaction case reports, the criteria do allow generalization to other types of case reports, regardless of discipline or specialty. Venulet has listed three types of case reports:

1. Early, unreviewed reports
2. Case reports, reviewed
3. Multiple cases

Authors should carefully consider the information contained in the case reports they are writing. If the above criteria seem to have applicability to their reports, they should describe the strata where their respective case reports should fall and certainly should discuss the limitations of the information the reports contain.

2. Commentary

The commentary allows the writer to express well-documented opinions espousing a specific point of view relative to a topic of interest. The commentary is essentially an editorial submitted by someone other than an editor (although an editor certainly can submit commentaries or editorials). In the commentary the author describes the status quo with regard to a topic and either agrees or disagrees with it. This publication channel allows the writer to thoroughly document a specific point of view with literature citations and then voice an opinion relative to acceptance or rejection of the adoption of a particular point of view.

3. Position Paper

Position papers allow the writer to take a firm stand on a topic of concern. This type of publication would be considered analogous to a legal brief in that a specific point of view is strongly held, documented, and presented. Another term for a position paper may be a ''white paper''. These papers allow an individual, department, school, or association to take a stance on a controversial or noncontroversial topic of interest. Position papers are similar to commentaries, but are usually much longer and more detailed.

4. Research Report

When publishing is considered by most researchers, the research report is what initially comes to mind. Research reports allow the writer to share the fruits of research with other scientists and interested readers. Because of the volume of articles most individuals must read, it is crucial to express the rationale, means, findings, and interpretation of research in a suitable fashion. Research reports ideally should be broken down into the following four sections:

1. Introduction
2. Methodology
3. Results
4. Discussion

For a truly worthwhile paper to be published, each of these sections must express the thoughts of the writer adequately and succinctly.

a. Introduction

The introduction is the section of a paper where the author describes the rationale for doing the study. Prior to presenting the rationale for the present study, a synopsis of what has been done by others in the same field must be given. This literature review should be sufficiently lengthy to acquaint the uninformed reader with the current status quo pertinent to the study as well as describe what has been done in the recent past that is relevant to the study. Readers familiar with the area of study will be scrutinizing this portion of the paper to ascertain that the writer in effect knows the area of research. This review of the research portion of the introduction should not be exceedingly lengthy; otherwise, it will detract from the paper as well as the reader's interest.

After the completion of the literature review component of the section, the author must present the rationale for the study in question. The author may base the need for doing the study on gaps that appear in the literature. In addition, the study may be predicated on the previous work he or she has completed in the field. If the author has completed previous works, these should be referenced in the literature review and introduction. This allows the reader to examine former work and will more adequately prepare the reader to examine the present study. Authors may assume that everyone in the discipline is familiar with their work; this is not always the case. Authors have so many avenues by which they must publish; readers who are not familiar with the previous papers need to examine these other sources to competently evaluate the present research or, perhaps, determine what else needs to be accomplished from a research standpoint.

A word of caution about computerized literature searches should be noted. Researchers should not rely solely on computerized literature searches in reviewing the literature. There are limits on what journals are abstracted and included in computerized searches. In addition, there can always be errors of omission of certain journals. Also, even the best abstracting publication is dependent on human data entry and key word entry by the author on the paper itself. The key word items required by many journals may be considered bothersome by authors or, worse yet, as unimportant. It has been suggested that authors need to structure their abstracts so that key aspects of purpose, methods, and results are reported with a partly controlled vocabulary and in a standardized format. This would assist readers in better evaluating the literature for applicability to their needs.[4]

b. Methodology

The methodology section is the component of the research write-up where the author(s) sufficiently present the research plan for review by the reader. This section should allow the reader to fully understand everything that was accomplished to carry out the research project. Put simply, this section should address a series of questions:

- What?
- Why?
- Where?
- When?
- Who?
- By whom?

If a reader can peruse the methodology section and have these questions answered, the section should be considered to have been adequately written. Conversely, if the reader is confused about the details of the study because of gaps in the answers to the above-mentioned questions, the section is flawed and should be revised.

In a journal article it may appear impossible to elaborate on every detail. However, to fully inform the reader of the whys and wherefores of the research, the writer must be instructive and complete in describing the research plan. The methodology section should not diminish the reader's attention, but insteady fully lay out what was to be accomplished by the research design. It should be noted at this point that, although the methods section is an important component of a research write-up, other components of the manuscript should not suffer because of undue attention focused on the methods section. Just as the methodology of a study should not drive a research study, the methodology section of a manuscript should not drive the writing of the paper describing the study.

c. Results

The results section of the manuscript is the location for the author to describe the findings of the research study. At this point in the manuscript the reader should not find any surprises relative to the presentation of findings and the reported results. The results should be detailed enough to allow the reader to follow the findings of the research, yet not belabor any particular finding.

Tables are an excellent medium for presentation of statistics that would be difficult to present in any other fashion. Tables should be titled appropriately and should flow smoothly. Turabian[5] has presented a detailed list for the appropriate listing of data in tabular form. If material presented in a table is self-explanatory and understandable without further amplification, it is unnecessary to explain the contents of the table in the text. In fact, it is not only unnecessary, but also superfluous. If reference can be made to a table and if the contents of the table are self-explanatory, the author has done all that can be expected. The presentation of research results should flow smoothly and allow for a suitable entry into the discussion section.

d. Discussion

The discussion section is the appropriate place for the author to tie the entire article into one package. The presentation of the discussion allows the author and the reader to follow the research results found in the present study with those from previous research, if applicable. If significant findings have been produced, placing them in context with the past research is a vital component of the manuscript. The discussion section must flow from the findings of the research project. It is inappropriate to conjecture during the discussion on what might have been found in the research. The author must discuss what *was* found and not dwell on what should have been or could have been found. Authors must discuss the results conservatively in the context of what was initially put in place to study by the research design. The discussion section must be complimentary to the preceding sections and serve to fully inform the reader of the interpretation of the author of the research findings. There should be no new findings presented in the discussion section. Any and all findings from a statistical standpoint (either descriptive or inferential) should have preceded any material presented in the discussion section.

5. Other Types of Reports

Other types of reports can emanate from other, varied research methodologies. For example, qualitative research is an area of investigation with a definite focus and specific outcome expectations. However, evaluation research requires different end points than those used for experimental or quasi-experimental research studies. The guidelines and expectations for publishing qualitative research must be realized by both editors and researchers. Smith[6] has noted that qualitative research contains four types of approach:

1. Interpretive
2. Artistic
3. Systematic
4. Theory driven

There are also different idealogies which exist within the qualitative research discipline.

III. PUBLICATION

A. HISTORICAL PERSPECTIVE

Scientific publication began in 1665 with the publication of the *Philosophical Transactions of the Royal Society of London* in England and the *Journal des Scavans* in France.[7] After these and many succeeding "journals" were successful, the academic publication business flourished. Prior to the publication of scholarly works in journals, scientists exchanged research progress through books, pamphlets, or treatises passed from one colleague to another.

B. EXPLOSION IN THE NUMBER OF PUBLICATIONS

Since these rudimentary beginnings the art and science of publication has exploded. There is now such a plethora of publication possibilities that it is becoming increasingly difficult to focus on a single publishing mode. It is also becoming extremely difficult to monitor the literature in one's own field of expertise. There is a virtual explosion in the number of journals, including specialty and subspecialty publications. Estimates of the number of journals range from 30,000 to 90,000.[7] Certainly there is a spectrum contained in these journals with respect to the circulation, scope of material, discipline(s) covered, etc. Nevertheless, the sheer number of publication possibilities is difficult to comprehend.

C. TYPES OF JOURNALS

Journals traditionally have been divided into scientific and professional classifications. The scientific journal is a refereed (peer-reviewed) journal that may be general in nature (e.g., *Journal of the American Public Health Association)* with a broad readership and contributor base. A refereed journal indicates that one or more reviewers have critiqued the manuscript in question. Ethically, the author's identity should not be known by the reviewer, and the identity of the reviewer should be unknown by the author of the manuscript. Journal editors work very diligently to maintain the confidentiality of both parties.

Scientific journals can be much narrower in scope from a professional point of view (e.g., *The Journal of the American Medical Association* or the *American Journal of Hospital Pharmacy*). The scientific journals may contain sections providing information, such as listing the dates for upcoming meetings. For the most part, readers and researchers see the scientific journal as a medium to exchange information regarding research.

D. PROFESSIONAL JOURNALS

Professional journals or publications may contain sections that are peer reviewed, but more often than not submissions are reviewed through an in-house review system or through critique by one or more members of a selected editorial board. These journals may provide an opportunity to publish a component of a research study, but should not be seen as an avenue to reach the scientific community. They do provide an opportunity to reach practitioners who in some cases are not a component of the scientific community. These comments should not be taken in a pejorative fashion; professional non-peer-reviewed publications provide a major service to professionals and scientists alike. The mission of the professional journal is simply different than that of a scientific journal.

E. PUBLISHING POSSIBILITIES

In order for researchers to be optimally productive they should determine the best publication source for their material. This exercise should not be seen as a limiting procedure, but as a focusing endeavor. Through the examination of one's expertise, a suitable audience, and publication possibilities the researcher can be better equipped to progress through the research milieu.

F. UNDERSTANDING PUBLISHING

1. The Process of Publishing

Since more often than not academic success is measured by publication prowess, researchers must understand the process of publication. The write-up and subsequent publication of research findings may be the most challenging and frustrating components of the research milieu. Perseverance and a thick skin are helpful attributes at this stage of research.

A useful component in the writing stage of research is the consideration of who the audience for the manuscript will be. If a particular journal is a possibility, it may be useful to examine the type of articles contained in the journal. Is there a specific outline used, referencing style, or other requirement specific to the journal? Consideration of these factors will allow the researcher to obtain a thematic sense of the direction of the editorial content of the journal. Obviously, if a journal contains articles from eclectic sources with varied contents, this approach may not be as useful.

2. Revising Papers

In most cases, the researcher must satisfy self-initiated questions as well as the questions of the editor and one or more reviewers. Often it is frustrating to meet one's own standards only to be informed that others do not share enthusiasm for the work. At the point of conflict between an author's view and that of the reviewer, the author must do what is felt to be best. In other words, the suggested changes and corrections the author feels comfortable with should be made and the article resubmitted for consideration. At this point, the editor may accept the manuscript without further changes or perhaps select several other reviewers to evaluate the manuscript. If changes can be made without altering the intended meaning or tone of the paper, the end result of a published paper is worth the added time and effort expended.

In cases of irreconcilable differences between the author and reviewer, the final decision will rest with the editor. Bishop[8] has suggested the role of an editor in these instances: "...the editor should be able to see and negotiate a compromise when an author has accepted the substance of a referee's criticism without necessarily having followed the recommendations to the letter."

3. Submission of Rejected Articles Elsewhere

If a manuscript is rejected outright, the author has the prerogative to submit the article for publication consideration elsewhere. It is not inappropriate for an author to consider several possibilities for publication after rejection. However, it is inappropriate for an author to submit duplicate manuscripts to two different journals at the same time. This deception is unethical and counterproductive to the sharing of research results in the scientific community. The phenomenon has not escaped the attention of journal editors. Some journals have blacklisted authors who were found to have made duplicate submissions; the editors refused to consider further papers from those authors for a period of 3 years.[8]

4. Plagiarism

Plagiarism is also another thorny issue which plagues authors, editors, and readers. Plagiarism can be defined as the intentional or unintentional transcription of another author's work. When intentional plagiarism occurs, the practice should be considered unethical and dealt with in the same fashion as is falsified documentation of research findings. The facts pertinent to the plagiarism must be exposed, apologies made, and retractions demanded. Unintentional plagiarism occurs when an author does not properly reference the use of other authors' original research. Garfield[10] has called this plagiarism "citation amnesia" and has suggested that it arises through oversight, forgetfulness, or lack of thoroughness on the part of authors. Authors need to be meticulous in their use of others' work in their own writing.

G. MULTIPLE PUBLICATION FROM RESEARCH STUDIES

On occasion a researcher will complete a research project that requires multiple publication of results because of the scope of the research project. If a major work contains enough substantive components, the author is justified in publishing several papers from the work. This publication of several components of the work separately is called for when publication of the entire amalgamation of components would be cumbersome.

For example, consider a hypothetical research project. A team of researchers has completed a study in which health care professionals have tested an investigational new smoking cessation drug in a patient population of smokers. The study utilized a double-blind crossover design. In the research project, health professionals counseled patients on the need for smoking cessation and sought to determine the efficacy of the smoking cessation drug. The authors of the study could write manuscripts focusing on the following varied aspects of the research project:

1. The results of the double blind crossover, placebo controlled comparison of the effectiveness of the investigational drug vs. the placebo drug
2. An examination of the health professionals' attitudes toward their role as smoking cessation counselors

3. Patients' attitudes toward the drug and/or health professionals' activities encouraging smoking cessation activities

Certain data would be shared by the three analyses resulting from examining each of the three items listed above. Demographic characteristics of the patients may be of interest in the first and third analyses, even though the actual type of analysis may vary. For example, item three would lend itself to some sort of multivariate analysis (i.e., multiple regression, path analysis, or LISREL). The analysis called for in item 1 may be analysis of variance or covariance, chi-square, or others. Each of these analyses conceivably could incorporate demographic data into the analysis. The remaining pieces of each subset of the data would not overlap. These distinct collations of the data will be of interest to differing and, in some cases, similar audiences. Through incorporating too much of the total research project in one manuscript the results of the project may be diluted. Because of this, each of the segments is worthy of a publication. Hence, researchers need to consider which components of projects could be written up as a research article. The incomplete reporting of research projects in an appropriate fashion is a disservice to the author, the research sponsor, and the scientific community.

H. FRAGMENTED PUBLICATION

Fragmented publication of overlapping results in different publications is not appropriate. The use of letters to the editor or other small-scale write-ups in order to beef up one's resume only serves to clutter the scientific literature. The above statement should not be interpreted as a criticism of letters to the editor sections of journals. Letters to the editor serve a useful purpose in communicating ideas in a propitious fashion. The point is that unwarranted publishing of overlapping components of a research project adds little to the scientific literature.

I. SOURCES OF HELP FOR WRITING

There are many fine sources of information for further help in answering questions relative to publishing and writing papers.[11-14] One attains success in writing and publishing by beginning with sound research questions and the perseverance to finish the research process by successfully publishing the results of the project. As with many other laborious processes in life, practice in writing allows the researcher to gain both confidence and skills. Another useful approach is to read what others write, both to critically examine their style and to evaluate how they progress from start to finish with their papers and reports. It is also helpful to have colleagues examine your research papers (an internal referee system) before submitting the manuscript for outside review. In any case, worthwhile efforts in writing and publishing lead to an enormous sense of personal reward and satisfaction.

REFERENCES — SECTIONS I TO III

1. **Kerlinger, F. N.,** *Foundations of Behavioral Research,* 3rd ed., Holt, Rinehart & Winston, New York, 1986, chap. 9.
2. **Begg, C. and Berlin, J.,** Publication bias and dissemination of clinical research, *J. Natl. Cancer Inst.,* 81(2), 107, 1989.
3. **Venulet, J.,** Informativity of adverse drug reactions in medical publications, *Drug Inf. J.,* 19(3), 357, 1985.
4. Ad Hoc Working Group for Critical Appraisal of the Medical Literature, *Ann. Intern. Med.,* 106, 598, 1987.
5. **Turabian, K. A.,** *A Manual for Writers of Term Papers, Theses, and Dissertations,* 5th ed., University of Chicago Press, Chicago, 1987.
6. **Smith, M.,** Publishing qualitative research, *Am. Educ. Res. J.,* 24, 173, 1987.

7. **Bishop, C. T.,** *How to Edit a Scientific Journal,* ISI Press, Philadelphia, 1984, chap. 1.
8. **Bishop, C. T.,** *How to Edit a Scientific Journal,* ISI Press, Philadelphia, 1984, chap. 2.
9. **Christiansen, D.,** Spectral lines. Multiple exposure, *IEEE Spectrum,* 20, 31, 1983.
10. **Garfield, E.,** More on the ethics of scientific publication: abuses of authorship attribution and citation amnesia undermine the reward system of science, *Curr. Contents,* 30, 5, 1982.
11. **American Psychological Association,** *Publication Manual of the American Psychological Association,* 3rd ed., American Psychological Association, Washington, D.C., 1983.
12. **Campbell, W. G.,** *Form and Style: Theses, Reports, Term Papers,* 7th ed., Houghton Mifflin, Boston, 1986.
13. **Schmid, C.,** *Statistical Graphics: Design Principles and Practices,* John Wiley & Sons, New York, 1983.
14. **Strunk, W. and White, E.,** *The Elements of Style,* 3rd ed., Macmillan, New York, 1979.

IV. SAMPLES OF REPORTED RESEARCH FINDINGS

The results of this research have been generated using the survey questionnaire discussed in Chapter 3, Section II. This is an example of reporting a research finding for possible publication.

A. CHOICE OF WESTERN VS. NATIVE HEALTH CARE PROVIDERS FOR SPECIFIC HEALTH PROBLEMS IN NIGERIA

ABSTRACT

In a pluralistic health care arena, individuals must choose from various systems of health care and Nigeria is no exception. The continued success of native medicine has been attributed to both patient and healer sharing similar beliefs on disease etiologies.

This study was therefore developed to identify some specific health problems for which respondents would utilize native rather than Western health care providers, and vice versa, in Nigeria. Six hundred and eighty adults were interviewed, using a survey questionnaire. Over fifty percent of the respondents would utilize the services of the native doctor for psychiatric care, broken bones, snake and insect bite, and convulsive disorders. However, the respondents would almost exclusively utilize Western medicine for prenatal care and childbirth, chicken pox, fainting, leprosy, and intestinal worms. Where a disease or illness has a cultural meaning, the use of native medicine would be more likely than the use of Western medicine.

Introduction

Nigeria, with a population of about 100 million, is comprised of various ethnic groups with the diversified cultural mix found in most developing nations.

The term native medicine should not imply that systems designated as such are stagnant replicas of those occurring thousands of years ago. It has been emphasized, for example, that Malaysian Chinese medicine today, although firmly rooted in tradition, is also a modern, innovative, changing system [1]. Each system, and the interrelationship of systems, reflects the contemporary mileau in which they function. Researchers interested in health and illness in a cross-cultural perspective are aware that the great majority of people do have recourse to non-medical healers. Various proposals have been made to account for the success and failures of such practitioners. The general consensus of opinion, however, seems to be that both patient and healer share similar beliefs [2]. The success or failure of healing was, therefore, attributed to the shared cognitive world in which they both live and function. Dobkin in her study indicated that not only do both healer and patient share similar beliefs about disease etiology, but they also both define illness within similar rubrics, even though the healer and patient may not participate equally in healing rituals or equally comprehend healing doctrines [3]. While Western medicine can produce a cure in the patient's mind, it does not remove the basic cause of his illness, such as a curse from God. Native healers, on the other hand, address such basic causes and are well poised to maximize psychological and cultural treatment of the illness.

In a pluralistic health care arena, individuals must choose from various treatment modalities available to them as a means to seek treatment of illness and to combat disease. In developing nations where Western systems of health care have only recently been added to the native system, the question of choice becomes very significant and Nigeria is no exception. People, therefore, have to choose from at least two distinct health care providers, each with a defined medical knowledge and practice in health related problems. The impact of multiple systems of health care services remains a little understood area. The plurality of health care enables one to switch from one type of practitioner to another in search of the best, cheapest, or most accessible means to care for health problems. The continued existence of medical pleuralism shows that no system holds a monopoly on healing. People's confidence in the efficacy of various healing practices is often attributable to the particular ways in which they conceptualize sickness etiologies rather than to the physical consequences of medical interventions.

Healing is anchored in particular social and cultural contexts. Each society has its own beliefs about illness, choices of treatment alternatives, sickness roles, and health care related institutions organized as a health care system. Healing is viewed differently across cultures and in different sectors of health care as well as between practitioners and patients. From the perspective of health care and its clinical functions, healing is the sum of the activities of the entire system of health care excluding prevention. In most developing nations, the introduction of Western medicine into a community affects its exclusive dependence on native medicine. Various studies have indicated that people in developing nations seek out Western practitioners for health problems in which manageable illness components dominate over unmanageable components [4]. One cannot properly conclude that traditional practices persist only because people believe they are efficacious. Efficacious practices persist in spite of undesirable outcomes. The undesirable outcomes are usually explained away to patients by invoking special circumstances surrounding the health problem. It has been argued that in healing the structure of the healing process itself is as important as any pharmacological medication [5—7]. Treatment provided by a native medical system may be particularly effective not only because of the efficacy of the medications used, but because of the character of the treatment itself. The method of treatment, therefore, becomes the potent active ingredient in healing.

Often people tend to think in the either/or category, in that if a person uses Western medicine then he/she cannot therefore also believe in and use native medicine. However, people can and do. This is true in many parts of the world including the United States [8—9]. The success of native medicine in filling certain needs of their clients illustrates the inadequacies in the Western health care system. Patients realize that both native and Western systems of health care have imperfections and as a result will switch from one system to another. Calculation of the relative efficacy of cures in both systems of health care is complicated when people utilize native medicine first and then turn to Western medicine or vice versa. There is some scientific evidence of mind and body inseparably contributing to the healing process [10]. A Western medical clinic or hospital which operates on a set of cultural assumptions different from that of its village patients might implicitly, though unwittingly, challenge their world view without being able to replace it. Western/cosmopolitan medicine, though provided by natives themselves, might reflect a too Western approach sometimes alien to patients elsewhere [11]. The prevalence of traditional medicine may reflect the prevalence of a different world view than that of the urban-oriented Western health practitioners [12—15].

A critique of Western practice, particularly when using the perspective of traditional practice to facilitate the critique, does not deny the tremendous achievements of Western medicine and the benefits derived from its use. But because Western medicine has forfeited the effective and social aspects of healing which were part of its heritage, emphasized Virchow [16], patients are often inhibited from taking advantage of its undeniable positive capabilities and for the same reason are prompted to rely on other practices of questionable effectiveness [17].

It may be assumed that the healing systems now existing in most developing nations will persist and certain traditional medical practices may gain recognition. The creation of an Association of Nigerian Traditional Herbal Medicine signals changes within that system as well as in how that system will be perceived by others. Apart from whatever contact may take place between Western and native providers, there are indications that changes may be expected within the national health care system itself, with greater emphasis on rural health care, the use of auxiliary personnel, and a team approach to primary health care services. International and national developments point to a closer rather than a more distant relationship between the various systems [18—19]. Lock in her paper indicated that although the dominant health care in Japan is Western, there exist traditional practitioners who are flourishing [20].

It has been assumed that indigenous healers primarily treat acute, self-limited (naturally remitting) disease; non-life-threatening, chronic diseases in which management of the illness is a larger component of clinical management than biomedical treatment of the disease; and secondary somatic manifestations of minor psychological disorders and interpersonal problems. In Taiwan people tend to go to Western doctors for the control of potentially life-threatening diseases, diseases which Western medicine is particularly effective in treating, and to indigenous/native healers for personal and culturally meaningful treatment of illnesses. They do not go to Western doctors for the treatment of fright, which is a cultural illness, but to healers poised to apply the culturally sanctioned treatment for this illness [1, 23]. This is not to say that indigenous therapy is entirely ineffective against all severe, acute, and life-threatening diseases. In fact, it may exert a placebo effect or directly affect such diseases. However, it has been known that all forms of psychotherapy seem to work for a wide range of mental disorders [21]. The universal therapeutic components of psychotherapy seem to be present in both modern and traditional forms [22]. Thus, there is much in favor of the argument that indigenous forms of psychotherapy may effectively treat certain psychological and perhaps also psychophysiological diseases. Moreoever, research on the physiological effects of meditation, behavior therapies, biofeedback, and placebos suggests a number of different ways by which indigenous therapies may affect biologically based diseases [23]. Since the epidemiological web causing and sustaining physiological diseases not infrequently includes major psychosocial factors, indigenous healing may at times work to affect such diseases by altering those factors.

However, at the same time, Western medicine's rapid and successful action against many acute diseases may also increase people's confidence in (or at least their reliance on) other traditional cures. For instance, Djurfeldt and Lindberg [24], in their research paper on South India, describe how practitioners of Western medicine tend to siphon off many of the sickness episodes against which traditional healers are least successful; that is, episodes which have become severe and have the poorest prognosis are generally taken to practitioners of Western medicine. Gould [25] also wrote on the disinclination of Indian laymen to seek out Western practitioners for ailments in which manageable illness components dominate over unmanageable components. One consequence has been that the differential rate of remissions is now more favorable to indigenous native medicine than to Western medicine.

Nichter [26] in his paper describes the tendency of many people to give credit to Western medicine for suppressing sickness symptoms and to traditional medicine for eliminating the cause of the sickness.

Realizing that in a pluralistic health care environment people must choose one health care system over another, this study was therefore developed to identify some specific illnesses, symptoms, and diseases to which respondents would utilize native rather than Western health care providers and vice versa.

Study Sample and Methodology

The Hausa, Ibo, and Yoruba tribes constitute about seventy-five percent of the Nigerian population and as a result, respondents from these tribes were surveyed. Six hundred and eighty people from these tribes were surveyed because, according to Sheatsley, the optimum sample for most surveys should be between two hundred and two thousand cases [27]. Samples less than two hundred were considered useful only in exploratory research, while samples over two thousand generally produce small increments in precision considering limited resources.

The study sample consists of relatively equal numbers of adult males and females (21 years or older). About one third of the six hundred and eighty people in the sample came from each of the three tribes. For the Hausa, Ibo, and Yoruba tribes residents in Maiduguri, Anambara, and Abeokuta areas were surveyed, respectively. A relatively equal number of the respondents living in the urban areas were studied. The cities and rural areas were identified in the nineteen Nigerian states. Those states not representative of the tribes of research interest were eliminated to insure a representative sample of each tribe.

The survey approach was used because it offered the advantage of wide scope and the study of large populations by selecting and studying samples chosen from the populations to discover the relative incidence, distribution and interrelations of sociological and psychological variables. Since survey research focuses on people, their beliefs, opinions, attitudes, motivations and behavior, it was well suited for this research study [28—29].

Results and Discussion

Using the survey questionnaire, respondents were asked to indicate the health care provider of choice for some specific illnesses, symptoms, and diseases.

For some specific health or medical problems, respondents seemed to exclusively prefer either a native or a Western health care provider. For broken bones, for example, about 66 percent of all the 680 people surveyed would first utilize native doctors for care. Other health-related problems to which respondents would also go first to the native doctors for care include snake or insect bites (59.3 percent), psychiatric problems (68.2 percent), and convulsions (57.8 percent). However, for some other health-related problems, such as a big cut from a knife or an axe, about 90 percent of the respondents would utilize Western health care. Other medical and health problems for which respondents would almost exclusively use Western health care were childbirth (88.1 percent), prenatal care (89.0 percent), chicken pox (88.2 percent), fainting (84.1 percent), leprosy (83.4 percent), and worms (80.0 percent), as indicated in Table I.

Respondents' choice of health care for these health related problems is a function of respondents' perceived severity and etiology of the illness, perceived effectiveness of the health care provider, cost of health care, and distance to be traveled to obtain care. Although most of the respondents would utilize the services of Western health care in serious life-threatening health problems, one can speculate that individuals would utilize native doctors for culturally meaningful treatment of diseases such as convulsions which may also be life threatening. However, to conclude that individuals used native medicine only for acute, non-life-threatening problems would be erroneous. Rather a more appropriate conclusion is

TABLE I
Respondent's Primary Choice of Health Care for Certain Health Problems (Illness, Symptoms, Disease)

Health Problem	Native Doctor f (%)	Hospital/Clinic f (%)	Other f (%)
Psychiatric	464(68.2)	206(30.3)	10(1.5)
Broken bone	448(65.9)	221(32.5)	11(1.6)
Snake or insect bite	403(59.3)	267(39.3)	10(1.5)
Convulsion	393(57.8)	277(40.7)	10(1.5)
Infertility	298(43.8)	358(52.6)	24(3.5)
Malaria	260(38.2)	395(58.1)	25(3.7)
Swollen legs, hands, etc.,	225(37.5)	406(66.6)	30(4.4)
Persistent headache	197(29.0)	453(66.6)	30(4.4)
Stomach problems	167(24.6)	484(71.2)	29(4.3)
High fever	165(24.3)	478(70.8)	37(5.4)
Sudden sickness	151(22.2)	504(74.1)	25(3.7)
Excessive tiredness	145(21.3)	480(70.6)	55(8.1)
Chest pain	133(19.6)	522(76.8)	25(3.7)
Bodily aches and pains	122(17.9)	509(74.9)	49(7.2)
Diarrhea/dysentery	119(17.5)	516(75.0)	54(6.6)
Worms	95(14.0)	544(80.0)	41(6.0)
Leprosy	93(13.7)	567(83.4)	20(2.9)
Fainting	84(12.4)	572(84.1)	24(3.5)
Chicken pox	64(9.4)	600(88.2)	16(2.4)
Childbirth	64(9.4)	599(88.1)	17(3.5)
Prenatal care	59(8.7)	605(89.0)	16(2.4)
Big cut from knife, axe, etc.	46(6.8)	614(90.3)	20(2.9)

one which realizes that where a disease has a cultural meaning let's say to its etiology, respondents would more often than not utilize native medicine and where that proves to be ineffective would then turn to Western medicine. For instance, most respondents would first utilize the services of native doctors for convulsions, as already indicated. Most health care providers would agree that convulsions are certainly neither a self limiting disease nor a non-life-threatening disease. Yet respondents prefer to use native medicine for its control. To these respondents, convulsions may signify a manifestation of a supernatural phenomenon and, as a result, inappropriate for Western medicine to deal with effectively. As for psychiatric problems, it has already been indicated that therapeutic components of psychotherapy seem to be present in both Western and native health care [22]. Research on placebos has also suggested a number of different ways by which indigenous therapies may affect biologically based diseases [23]. For other health problems, such as broken bones or snake and insect bites, the choice of native health care can be due to the native doctors' effectiveness in dealing with these problems based on respondents' past experiences.

Respondents used Western health care providers for most of the illnesses, symptoms, and diseases, and this is a function of their proven effectiveness. As a result, Western medicine is seen as being the most appropriate source of health care for these health-related problems.

REFERENCES

1. **Dunn, F. L.:** Medical Care in Chinese Community in Peninsular Malaysia in Medicine in Chinese Cultures — Comparative Studies on Health Care in Chinese and Other Societies, edited by Kleinman et al., Dept. of Hlth. Educ. and Welfare. Publ. No. (NIH) 75-683, NIH, U.S. Pub. Health Service, Washington, D.C., 1975.

2. **Enrewald, J.:** Psychotherapy: Myth or Method? Academic Press, New York, 1958.

3. **Dobkin Derios, M.:** Socio-economic Characteristics of and Amazon Urban Healer's Clientele. Soc. Sci. Med. 15B, Jan. 1981.

4. **Gould, H.:** The Implications of Technological Change for Folk and Scientific Medicine. Am. Anthropol. 59, 3, June, 1957.

5. **Fernaudex, J.:** Persuasions and Performances of the Beast in Everybody and the Metaphors of Every Man. Daedalus 101, 39, 1971.

6. **Kleinman, A. M.:** Medicine's Symbolic Reality. Inquiry 16, 206, 1973.

7. **Moerman, D. E.:** Anthropology of Symbolic Healing. Curr. Anthropol. 20, 59, 1979.

8. **Wolff, R. J.:** Modern Medicine and Traditional Culture Confrontation on the Malay Peninsula. Human Organ. Clgh. Bull. 24, 339, 1965.

9. **Singer, P.:** Traditional Healing: New Science or New Colonialism: Conch Mag. Ltd., New York, 1977.

10. **Pelletier, K. R.:** Mind as Healer, Mind as Slayer — A Holistic Approach to Preventing Stress Disorders. Delacorte Press, New York, 1977.

11. **Doyal, L. and Pennel, I.:** Pox Britanica, Health, Medicine and Underdevelopment, Race and Class 18, 155, 1976.

12. **Haggenhanger, H. K.:** Hospital Assistant in Malaysian Rural Health Care. Med. J. Malaysia 33, 165, 1978.

13. **Rappapont, H.:** The Tenacity of Folk Psychotherapy: A Functional Interpretation. Soc. Psychiatry 12, 127, 1977.

14. **Jones, W. T.:** World View and Asian Medical Systems: Some Suggestions for Further Study. In Asian Medical Systems, edited by Leslie, C. P 383—404, University of California Press, Berkeley, 1976.

15. **Torrey, E. F.:** The Mind Game: Witchdoctors and Psychiatrists. Bantam Books, New York, 1973.

16. **Acherknecht, E. H.:** Rudolf Virchow: Doctor, Statesman, Anthropoligist. University of Wisconsin Press, Madison, Wisconsin, 1953.

17. **Singer, P.:** Traditional Healing: New Science or New Colonialism. Conch Mag. Ltd., New York, 1977.

18. **Taylor, C. E.:** Asian Medical Systems, A Symposium on the Role of Comparative Sociology in Improving Health Care. Soc. Sci. Med. 7, 307, 1973.

19. **Adumwagun, Z.:** The Relevance of Yoruba Medicine Men in Public Health Practice in Nigeria. Public Health Rep. 84, 1085, 1969.

20. **Lock, M.:** The Organization and Practice of East Asian Medicine in Japan. Continuity and Change. Soc. Sci. Med. Men. Anthropol. 14B, 245, Nov. 1980.

21. **Luborsky, L.:** Comparative Studies of Psychotherapy: Is it Time that Everyone has Won and All Have Prizes? Arch. Gen. Psychiatry 32, 995, 1975.

22. **Frank, J.:** Therapeutic Components of Psychiotherapy. J. Nerv. Ment. Dis. 159, 325, 1974.

23. **Tseng, W.:** Traditional and Modern Psychiatric Care in Taiwan, in Culture and Healing in Asian Societies, Kleinman et al. Eds., Schenkman, Cambridge, 1978.

24. **Djurfeldt, G. and Lindberg, S.:** Pills Against Poverty: A Study of the Introduction of Western Medicine into a Tami Village. Student Literature, Lund, 1975.

25. **Gould, H.:** The Implication of Technological Change for Folk and Scientific Medicine. Am. Anthropol. 59, 3, 507 June 1957.

26. **Nichter, M.:** Patterns of Resort in the Use of Therapy Systems and Their Significance for Health Planning in South Asia. Med. Anthropol. 2, 29, Spring 1978.

27. **Sheatsley, P. B.:** Survey Design, in Robert Ferber (Ed.), Handbook of Marketing Research. McGraw-Hill, New York, 1974, 2.

28. **Dean, J. P., Elchhon, R. L., and Dean, L. R.:** The Survey, An Introduction to Social Research. Appleton-Century Crofts, New York, 247.

29. **Babbie, E. R.:** Survey Research Methods, Wadsworth Publishing, Belmont, California, 1973.

B. NIGERIAN'S USE OF NATIVE AND WESTERN MEDICINE FOR THE SAME ILLNESS*

Nigerians' Use of Native and Western Medicine for the Same Illness

EUCHARIA E. NNADI, PhD
HUGH F. KABAT, PhD

Dr. Nnadi is assistant professor, College of Pharmacy, Florida Agricultural and Mechanical University, Tallahassee, Fla. 32307. Dr. Kabat is associate dean for academic affairs and professor, College of Pharmacy, University of Minnesota, Minneapolis. Tearsheet requests to Dr. Nnadi.

SYNOPSIS

The impact of dual systems of health care on those who use them is not well understood. A study was undertaken in Nigeria to study this impact. A questionnaire about use of dual health care systems was prepared and tested and then administered to a study sample. This sample included representatives of the Hausa, Ibo, and Yoruba tribes, urban and rural residents, and Christians, Muslims, and traditional worshippers.

Among 680 respondents, more of the Muslims than the other religious groups had used both Western and native medical care systems for the same illness, and the relationship between such dual use and religious preference was found to be significant. Also, almost three-fourths of the respondents who had used both native and Western medicine for the same illness had done so more than once. Among respondents who had used both for the same illness many times, there were almost twice as many males as females. Among the Ibos, 81.9 percent had used both systems more than once, compared with 77.2 percent of the Yorubas and 62.2 percent of the Hausas.

The existence of pluralistic medicine is fairly common in both developing and developed countries. However, switching back and forth from native to Western medicine makes determinations of the relative efficacy of cures difficult. Also, simultaneous use of drugs from both native and Western health care providers has the potential of harming or even killing the patient.

NATIVE MEDICINE CAN BE DEFINED as a system of medicine that is traditional, nonconventional, or non-Western. Dunn, for example, has emphasized that Malaysian Chinese medicine today, although firmly rooted in tradition, is a modern, innovative, changing system (*1*). Each health care system, as

* From *Public Health Reports*, 99(1), 93, 1984.

well as the interrelationship of traditional and Western systems, reflects the contemporary milieu in which it functions.

Researchers interested in health and illness from a cross-cultural perspective are aware that the great majority of people in various parts of the world have non-Western healers available. Different explanations have been offered for the success and failure of such practitioners. The consensus of opinion, however, seems to be that both the patient and the healer share similar beliefs (*2*). The success or failure of healing is, therefore, attributed to the shared cognitive world in which both the patient and the healer live and function. Dobkin has indicated that not only do both healer and patient share similar beliefs about disease etiology, but they also define illness within similar rubrics, even though the healer and the patient may not participate equally in healing rituals or equally comprehend healing doctrines (*3*). Native healers, in addition, are in a position to make the most of the psychological and cultural treatment of illness.

In a pluralistic health care arena, patients choose from the various treatment modalities available for treating illness and combating disease. In developing nations where the Western system of health care has only recently been added to the native system, the question of choice becomes extremely significant, and Nigeria is no exception. People have to choose from at least two distinct health care providers, each with defined medical knowledge and practices in health-related problems.

The impact of multiple systems of health care remains a little understood area. Plurality of health care enables a person to switch from one type of practitioner to another in search of the best, cheapest, or most accessible means of care. The continued existence of medical pluralism shows that no system holds a monopoly on healing. People's confidence in the efficacy of various healing practices is often due to the particular ways in which they perceive the etiologies of sicknesses rather than to the physical consequences of medical interventions.

Healing is anchored in particular social and cultural contexts. Each society has its own beliefs about illness, the choice of treatment alternatives, sick roles, and the health-care-related institutions that are organized as a health care system. Healing is viewed differently across cultures, in different sectors of health care, and by practitioners and patients. From the perspective of the health care system and its clinical functions, healing is the sum of the activities of the entire system of health care; prevention is excluded.

In most developing nations, the introduction of Western medicine into a community reduces the exclusive dependence on native medicine. Various studies have indicated that people in developing nations seek out Western practitioners for those health problems in which manageable illness components dominate over unmanageable components. (*4*).

One cannot conclude that traditional practices persist only because people believe they are efficacious, since such practices persist in spite of undesirable outcomes. Undesirable outcomes are usually explained by suggesting that special circumstances surround the health problem.

In healing, it has been argued that the structure of the health process itself is as important as any pharmacological medication (*5–7*). Treatment provided by a native medical system may be effective not only because of the efficacy of the medications used but because of the character of the treatment itself. The method of treatment, therefore, becomes the most active ingredient in healing.

Some believe that if a person uses Western medicine, he or she cannot also believe in and use native medicine. However, many people can and do so, in many parts of the world, including the United States (*8, 9*). The success of practitioners of native medicine in filling certain needs of their clients suggests that there are inadequacies in the Western health care system. Patients realize that both native and Western systems of health care have imperfections and, as a result, will switch from one system to another. Calculations of the relative effectiveness of curers in both systems is complicated when people use native medicine and then turn to Western medicine and vice versa. When both health care systems are used, the amount of time and money spent in seeking health care can be excessive.

Because the use of both health care systems by the same person is common, a study was designed to determine the extent of its occurrence in Nigeria and the characteristics of the persons engaging in such dual use.

Study Sample and Methods

In the study, 680 respondents from the Hausa, Ibo, and Yoruba tribes, which constitute 75 percent of Nigeria's approximately 100 million population, were surveyed. According to Sheatsley, the optimum sample for most surveys is between 200 and 2,000 persons (*10*). Samples of less than 200 are considered useful only in exploratory research; samples of more than 2,000 generally produce only small increments in precision, and the greater cost of surveying large samples must be considered when resources are limited.

The study sample consisted of relatively equal numbers of males and females 21 years or older. About one-third of the 680 people in the sample came from each of three tribes. Residents of the Maiduguri area represented the Hausa tribe, residents of the Anambara area the Ibos tribe, and residents of the Abeokuta area the Yoruba tribe. People from Abeokuta represent the Egba Yoruba well, but do not necessarily represent the people of other Yoruba kingdoms. Although residents of the city of Maiduguri are primarily members of the Kanuri tribe, there were sufficient Hausas living in the city to provide an adequate sample for the study.

The cities and rural areas in the 19 Nigerian States were identified. Approximately equal numbers of the study sample lived in urban and rural areas. About 63 percent of those surveyed were Christians, 20 percent Muslims, and 15 percent traditional worshippers. Table 1 summarizes the demographic characteristics of the study sample. Almost half (48.4 percent) of the respondents had only a primary education or less, 44 percent had a secondary education, and 7.5 percent had a college education. Almost three-fourths were in the low-income group, earning 200 Nigerian Naira (U.S. $300) or less a month. Of the remaining respondents, less than 10 percent were in the high-income group, that is, earning more than 400 Nigerian Naira (U.S. $600) a month. More than three-fourths were or had been married.

More than half (52.8 percent) of those surveyed were between 35 and 49 years of age; less than one-tenth were more than 49. Table 1 shows that 37 percent of the study sample were civil servants; the rest were farmers, traders, hospital employees, native doctors, housewives, and so forth.

The Hausas in the study sample were from the Bida rural area and the Maiduguri urban area, in Borno State: the Ibos were from the Ugbo rural area and the Enugu urban area, in Anambara State; and the Yorubas were from the Afebo rural area and the Abeotuta urban area, in Ogun State.

The major difficulty in interviewing respondents was accessibility. Nigeria's poor roads made transportation to the areas of the study sample, especially the rural areas, difficult. Since road maps were not easily available, the interviewers had to depend on getting directions from people who either had lived or had been in the areas targeted for the research.

A questionnaire for the planned survey was carefully constructed. Every effort was made to assure that the possible answers to the multiple-choice questions were exhaustive and mutually exclusive. Consideration was given to, but not limited to, the following factors: the clarity, length, unidimensionality, and specificity of the questions; whether the questions were socially embarrassing; and the questionnaire's readability.

Table 1. Demographic characteristics of the 680 respondents

Demographic characteristics	*Number*	*Percent*
Tribe:		
Hausa	226	33.3
Ibo	228	33.4
Yoruba	226	33.3
Sex:		
Male	341	50.1
Female	339	49.9
Religion:		
Christian	428	62.9
Muslim	137	20.2
Traditional worship	102	15.0
Others	13	1.9
Residence:		
Urban	332	48.8
Rural	348	51.2
Education:		
None or only primary	329	48.4
Secondary	300	44.1
College	51	7.5
Income per month in Nigerian Naira: [1]		
200	501	73.6
200 to 400	112	16.5
More than 400	67	9.9
Marital status:		
Single	149	22.9
Married	488	71.8
Divorced	9	1.3
Widow or widower	29	4.3
Separated	5	0.7
Age:		
21–34 years	359	52.8
35–49 years	260	38.2
More than 49 years	61	9.0
Occupation:		
Civil servant:		
Not teacher	173	25.4
Teacher	90	13.2
Not civil servant:		
Hospital-related work	58	8.5
Housewives	123	18.1
Native doctors	22	3.2
Farmers	84	12.4
Traders	82	12.1
Others	48	7.1

[1] 1 Nigerian Naira = $1.50 U.S.

Before its administration, the questionnaire was sent to U.S. and Nigerian sociologists, biometricians, and health care professionals and to Nigerian students for a critique and suggestions for improvement. In addition, it was pretested on a group of seven Nigerians (four men and three women), which included two health care professionals, two civil servants, a businessman, an illiterate housewife, and a youth corps volunteer from each of the tribes that was to be studied. After this group had critiqued the questionnaire, final revisions were made.

The survey instrument assessed demographic information, as well as attitudes, opinions, and behaviors. Multiple-choice questions were used because they require a minimum amount of the respondent's time (*11*). This type of questionnaire also facilitates data analysis. The questions were randomly located in the questionnaire except for those that could logically be grouped together. The instructions for the questionnaire were subjected to the Raygor test and found to be at a sixth-grade reading level (*12*).

A survey was used in the study because it offered the advantage of a wide scope and the opportunity to study large populations. The selection and study of large population samples make it possible to discover the relative incidence, distribution, and interrelations of sociological and psychological variables. A survey, because it can focus on people's beliefs, opinions, attitudes, motivations, and behavior, was well suited to our study (*13*).

The questionnaire was self-administered by all the survey participants who could read and write English. If they could not, the questionnaire was read to them in their native language. A disadvantage of the questionnaire method, especially when the instrument is administered by personal interview, is that a respondent may give what is perceived as a socially desirable response because a truthful response would be socially embarrassing. To minimize this tendency, respondents were informed before each interview of the confidentiality and anonymity of the study. Every effort was made to ensure that the respondents understood that no information that they gave would be used against them.

Results and Discussion

In addition to questions to elicit demographic characteristics, respondents were asked if they had ever used both Western and native medicine for the same illness. About half, that is, 312 or 45.9 percent, had used both types of medicine to care for the same illness; 368, or 54.1 percent, had not. Of the 312 who had used both types, 28.2 percent had done so for the same illness only once. However, the majority, once having used both systems during the course of a single health problem, tended to repeat the dual use rather frequently.

Those who used both Western and native medicine for the same illness were predominantly males, Muslims, Hausas, civil servants, urban residents, 21–34 years of age, of high income (that is, with 400 or more Nigerian Naira a month—1 Naira = \$1.50 U.S.), and had a secondary but not a college education. The chi square test, with 0.05 as the level of significance, was used to determine if a significant relationship existed between a demographic variable and the use of both systems of health care for the same illness.

Table 2 shows that the relationship between the respondent's sex and use of the two health care systems was statistically significant ($P < 0.05$). Although 47.5 percent of the men used both native and Western medicine, only 39.5 percent of the women did.

The relationship between dual use and religious preference was also found to be significant ($P < 0.01$). More of the Muslims than of the other religious groups used both systems of medicine for the same illness—59.1 percent, compared with 42.5 percent of Christians and 26.5 percent of traditional worshippers (people who did not belong to conventional religious groups but honored and offered sacrifices to their ancestors for protection).

There was no significant relationship between the respondent's occupation and use of both systems. A greater proportion of civil servants (46.8 percent) than those with other occupations (41.5 percent) used both systems for the same illness.

Among the three tribal groups surveyed, 55.8 percent of the Hausas, 43.8 percent of the Yorubas, and 31.1 percent of the Ibos had used both systems of health care during the course of the same illness. The relationship between being a member of one of these tribes and use of both care systems was significant ($P < 0.01$). Although 47.6 percent of the urban residents had used both systems for the same illness, only 39.7 percent of the rural residents had done so.

There was no significant relationship between the respondent's level of education and use of both systems. Nor was there a significant relationship between the respondent's income level and use of both systems. Respondents, regardless of income level or age, used Western and native medicines for the same illness in a similar way. The practice of using health care providers in both health care systems cuts across occupations, educational levels, income levels, and age groups.

As indicated earlier, 71.8 percent of the respondents who had used both native and Western medicine for the same illness had done so more than once. This repeated dual use is a cause for concern because the more frequently a person uses both native and Western medicine for the same illness, the greater the chance that he or she will experience undesirable and harmful drug interactions. Native doctors use various treatment modalities in caring for their patients, and some herbs used in their remedies can have pharmacological effects. Drugs provided a patient by both native and Western health care providers may, if used simultaneously, have synergistic or antagonistic effects. The use of drugs from both sources could thus result in harm to the patient or even death.

Further analysis was done to identify any relationship that might exist between the demographic variables of the respondents and the frequency with which they used both native and Western medicine for the same illness. The respondent's sex and tribal origin were found to be significantly related to the frequency of use of both Western and native medicine for the same illness. More than three-fourths (76.7 percent) of the men used both systems for the same illness more than once, compared with 65.7 percent of the women (table 3).

Of the Ibos who had used both systems, 81.9 percent had done so more than once, compared with 77.2 percent of the Yorubas and 62.2 percent of the Hausas who had used both systems. This result is interesting in that although more of the Hausas than the Ibos had engaged in dual usage, they had done so less frequently than the Ibos.

The relationship of the demographic variables of religion, occupation, residence, education, income, and age to the use of both types of health care for the same illness was not significant. Thus, although the respondents had different income levels, religious preferences, ages, occupations, and residences, there was no significant relationship between these factors

Table 2. Respondents' use of both native and Western health care for the same illness by sex, religion, occupation, tribe, residence, education, income, and age

Demographic characteristics	Did not use both systems		Used both systems		Chi square values [2]	P values
	Number	Percent	Number	Percent		
Sex:						
Male	179	52.5	162	47.5	4.03, 1 df	< 0.05
Female	205	60.5	134	39.5		
Religion: [1]						
Christian	246	57.5	182	52.5	45.81, 2 df	< .01
Muslim	56	40.9	81	59.1		
Traditional worship	75	73.5	27	26.5		
Occupation:						
Civil servant	140	53.2	123	46.8	1.62, 1 df	< .21
Not civil servant	244	58.5	173	41.5		
Tribe:						
Hausa	100	44.2	126	55.8	27.97, 2 df	< .01
Ibo	157	68.9	71	31.1		
Yoruba	127	56.2	99	43.8		
Residence:						
Urban	174	52.4	158	47.6	4.35, 1 df	< .05
Rural	210	60.3	138	39.7		
Education:						
None or only primary	191	58.1	138	41.9	1, 2 df	< .61
Secondary	163	54.3	137	45.7		
College	30	58.8	21	41.2		
Income:						
Low	287	57.3	214	42.7	2.31, 2 df	< .32
Middle	65	48.0	47	42.0		
High	32	47.8	35	52.2		
Age:						
21–34 years	193	53.8	166	46.2	2.27, 2 df	< .32
35–49 years	155	59.6	105	40.4		
More than 49 years	36	59.0	25	41.0		

[1] 13 respondents (less than 2 percent) belonged to other religious denominations.

[2] df = degrees of freedom.

Table 3. Frequency of respondents' use of both native and Western health care for the same illness by sex, religion, occupation, tribe, residence, and education

Demographic characteristics	*Once* Number	*Once* Percent	*A few times* Number	*A few times* Percent	*Many times* Number	*Many times* Percent	*Chi square values* [1]	*P values*
Sex:								
Male	40	23.3	101	58.7	31	18.0	6.31, 2 df	< 0.05
Female	48	34.3	77	55.0	15	10.7		
Religion:								
Christian	44	23.4	115	61.2	29	15.4	9.41, 4 df	< .06
Muslim	36	40.0	41	45.6	13	14.4		
Traditional worship	6	21.4	18	64.3	4	14.3		
Occupation:								
Civil servant	39	30.2	71	55.0	19	14.7	0.47, 2 df	< .79
Not civil servant	49	26.8	107	58.5	27	14.8		
Tribe:								
Hausa	51	37.8	58	43.0	26	19.2	19.82, 4 df	< .01
Ibo	13	18.1	50	69.4	9	12.5		
Yoruba	24	22.8	70	66.7	11	10.5		
Residence:								
Urban	44	26.8	95	57.9	25	15.2	0.33, 2 df	< .85
Rural	44	29.7	83	56.1	21	14.2		
Education:								
None or only primary	43	30.5	78	55.3	20	14.2	2.12, 4 df	< .72
Secondary	37	25.0	87	58.8	24	16.2		
College	8	34.8	13	56.5	2	8.7		

[1] df = degrees of freedom.

and the frequency with which both types of health care were used for the same illness.

Conclusion

The use of both Western and native medicine by the same person, and even for the same illness, was found to be common in Nigeria, just as it is in many other parts of the world. When both Western and native health care systems are available, people may switch back and forth from one type of practitioner to another or use both types simultaneously. Such dual use complicates the task of determining the efficacy of the treatment prescribed by either practitioner. Moreover, simultaneous use of medicine prescribed by a native practitioner and medicine prescribed by a Western practitioner could result in harmful drug interactions and adversely affect the user's health. Therefore, the study of dual or multiple use of health care systems represents an important area for continuous research.

References

1. Dunn, F. L.: Medical care in Chinese communities of Peninsular Malaysia. *In* Medicine in Chinese cultures—comparative studies on health care in Chinese and other societies, edited by A. Kleinman, P. Kunstadter, E. Alexander, and J. L. Gale. Department of Health, Education, and Welfare Publication No. (NIH) 75–653. Washington, D.C., 1975, pp. 297–326.
2. Enrewald, J.: Psychotherapy: myth or method? Academic Press, New York, 1958.
3. Dobkin DeRios, M.: Socio-economic characteristics of an Amazon urban healer's clientele. Soc Sci Med 15 (pt. B, Medical Anthropology): 51–63, January 1981.
4. Gould, H. A.: The implications of technological change for folk and scientific medicine. Am Anthropologist 59: 507–520, June 1957.
5. Fernaudex, J. W.: Persuasions and performances of the beast in everybody and the metaphors of everyman. Daedalus 101: 39–60 (1972).
6. Kleinman, A. M.: Medicine's symbolic reality. Inquiry 16: 206 (1973).
7. Moreman, D. E.: Anthropology of symbolic healing. Curr Anthropology 20: 59–80 (1979).
8. Wolff, R. J.: Modern medicine and traditional culture confrontation on the Malay Peninsula. Human Org 24: 339–345 (1965).
9. Singer, P., editor: Traditional healing: new science or new colonialism.. Essays in critique of medical anthropology. Conch Magazine, Ltd. (Publishers), Buffalo, N.Y., 1977.
10. Sheatsley, P. B.: Survey design. *In* Handbook of marketing research, edited by R. Ferber. McGraw-Hill, Inc., New York, 1974, pp. 2-66 to 2-89.
11. Manasse, H. R., and Randall, L. L.· Research methods. Am J Hosp Pharm 37: 694–700, May 1980.
12. Raygor, A. L.: The Raygor Readability Estimate: a quick and easy way to determine difficulty. Paper presented to the National Reading Conference, Atlanta, Ga., Dec. 2, 1976.
13. Dean, J. P., Elchhorn, R. L., and Dean, L. R.: The survey: an introduction to social research. Appleton-Century-Crofts, New York, 1967, 2d ed., p. 247.

INDEX

Q

R

S

T

U

V

W

Y

Z